GM CROPS

GM CROPS

THE IMPACT AND THE POTENTIAL

Jennifer A Thomson

CSIRO
PUBLISHING

National Library of Australia Cataloguing-in-Publication entry

Thomson, Jennifer A.
    GM Crops : the impact and the potential.
    Includes index.
    ISBN 978 0 64309 160 3.
    ISBN 0 643 09160 2.

    1. Crops – Genetic engineering. 2. Transgenic plants.
    I. Title.

    631.5233

Published exclusively in all territories of the World, excluding North and South America, by:

CSIRO PUBLISHING
150 Oxford Street (PO Box 1139)
Collingwood VIC 3066
Australia

Telephone: +61 3 9662 7666
Local call: 1300 788 000 (Australia only)
Fax: +61 3 9662 7555
Email: publishing.sales@csiro.au
Web site: www.publish.csiro.au

Published exclusively in North and South America, as *Seeds for the Future: The Impact of Genetically Modified Crops on the Environment,* with ISBN 978-0-8014-7368-5, by:

Cornell University Press
Web site: www.cornellpress.cornell.edu

Front cover photo by James Kelly
Set in Times New Roman PS and Stone Sans
Cover and text design by James Kelly
Typeset by Barry Cooke Publishing Services
Printed in Australia by Ligare

Foreword

This is an eagerly awaited sequel to Jennifer Thomson's book *Genes for Africa*. Like its predecessor it is written with the intelligent layperson in mind. It is about the frontiers of 21st century science yet the non-scientist will find that it is written in a lucid and straightforward manner. Any technical terms are clearly explained.

In *Genes for Africa* Jennifer Thomson argued the case for the potential benefits of genetically modified (GM) crops in Africa to help the continent eventually overcome the terrible scourge of hunger that afflicts so many of its inhabitants. Pests, disease and drought greatly reduce the yields of poor farmers and often the products of agriculture are not as nutritious as they could be. In other parts of the world experiments and, indeed, large-scale commercial cultivation have shown that GM crops can deliver crops that yield better and are more resilient.

Nevertheless there are potentially significant environmental hazards that have to be addressed and this is what Jennifer Thomson aims to do in the following pages.

In the first chapter she recapitulates the arguments of her first book, bringing them up to date. Then she goes on to examine the history of GM crops in both the developed and developing countries, recounting the stories and experiences of scientists, farmers and environmentalists. Jennifer Thomson examines Bt cotton: has it reduced the incidence of pesticide spraying and hence reduced the poisoning of farmers and farm workers? Does it damage soil insects? Do pests become resistant?

She asks similar kinds of questions of Bt maize and herbicide-resistant soybeans. Will these crops mean fewer sprays but still pose a potential risk to wildlife? She answers these and many other environmental questions with dispassionate care.

Two chapters look in general at the impact of GM crops on biodiversity and at the potential hazard of pollen spread from GM crops to other crops and to wild relatives.

The record to date on all these issues is encouraging. There have been few negative environmental impacts, and none of great consequence. If you produced a kind of score card, the environmental and health benefits would far outweigh the costs.

However, this does not mean we should not be vigilant. Each new GM crop needs to be thoroughly examined for its potential downsides. This means there have to be adequate biosafety regulations in place for both the developed and developing countries.

In her final chapter she lists some of the new GM developments in the pipeline – crops that produce life-saving drugs, including vaccines, plants that clean up the environment and plants with improved nutritional qualities, especially vitamin A precursors. These are to be welcomed for the benefits they will bring, but we need to keep constant watch on their likely environmental consequences. In the following pages you will find the issues that GM crops raise and the questions that have to be asked.

Professor Sir Gordon Conway KCMG FRS
Imperial College, London

Contents

Acknowledgements

I would never have embarked on writing a second book on genetically modified (GM) crops if it hadn't been for Professor Sir Gordon Conway's review, published in *Nature*, of my first book, *Genes for Africa: Genetically Modified Crops in the Developing World*, in which he stated that I could have expanded further on the environmental impacts of GM crops. I still might not have written it if I hadn't received a fellowship from the Rockefeller Foundation to spend three weeks working on it at the Villa Serbelloni in Bellagio, Italy, in July 2005. My grateful thanks to Pilar Palacia, our gracious host, and my marvellous co-fellows, most notably Roger Wilkins, who gave me such support during that time (and overlooked the fact that I did my laundry in the tumble-drier – twice). Then, to the many friends and colleagues who read various sections of the book and some of whom allowed me to use their photographs: Klaus Ammann, Bruce Chetty, Claude Fouquet, Dennis Gonsalves, Carl Pray and David Tribe among others. Grateful thanks go to Jennifer Eidelman, an outstanding librarian, who found every article I ever requested. Then to my IT 'guru' Nikki Campbell who cheerfully sorted out all my computer problems and managed the photographs. Finally I am indebted to geneticist Nancy van Schaik, who proofread and commented on the entire book and saved me from making some crashing errors.

Introduction

What impacts could genetically modified (GM) crops have on the environment? Could insect-resistant crops affect non-target insects? Could herbicide-resistant plants produce uncontrollable weeds? It is now clear that GM crops can increase yields but could they have an adverse effect on wildlife? Or could GM crops benefit the environment by reducing the spraying of harmful chemicals, and thus also reduce the impacts of spraying on human health? Could GM crops even reduce the levels of toxic chemicals in the soil?

Many of these questions are exercising the minds of those who are both for and against the use of GM crops. This book is therefore aimed at addressing, among others, questions such as:

- will insects develop resistance to the bacterial toxin produced by transgenic insect-resistant (Bt) crops?
- will the planting of herbicide-tolerant GM crops result in the development of a range of plants that are resistant to this herbicide (i.e. superweeds)?
- will insect-resistant crops lead to the death of non-target insects, hence decreasing biodiversity?
- will insect-resistant crops have negative effects on soil organisms?
- will GM crops cross-pollinate landraces and can this lead to their extinction?
- will GM crops lead to a decrease in plant biodiversity?

One comparison to keep in mind throughout this book is the one between GM crops and their conventionally bred counterparts. When the question of whether GM crops have created new problematic weeds is asked, a similar question needs to be posed with respect to their conventionally bred counterparts. Similarly the question of whether GM crops lead to a decrease in plant biodiversity needs to be asked also of their conventionally bred counterparts. Agriculture inevitably has an impact on the environment. The question therefore remains of what is a reasonable trade-off between crop production, wildlife and consumer concerns?

The difference between food production in the developed and the developing world needs to be recognised. I am an African and I therefore view food production and security from an African perspective. This is very different from a European perspective. In Europe, despite an overproduction of food, farmers are subsidised by their governments. It has been calculated that an average cow in Europe receives a subsidy of US$1 a day, more than the daily income on which an average African tries to survive.

Moreover crop yields are much lower in Africa than on many other continents. For instance the average yield among commercial farmers in South Africa is about three tonnes per hectare. Among small-scale farmers it could be as low as 0.1 tonnes per hectare. In the USA this could be as high as 10–15 tonnes per hectare.

The causes of poor agricultural production and resultant food insecurity in sub-Saharan Africa are many and complex. They include poor soil, harsh ecological conditions, poor infrastructure for transport of, and access to, food and declining investment in agricultural science. In sub-Saharan Africa agriculture provides about 70% of employment, 40% of exports and one-third of Gross National Profit (Conway and Toenniessen 2003). Two-thirds of the region's people live on small-scale, low-productivity farms. Often, the food a family can produce plus what they can afford to buy is not sufficient and as a result many Africans, mainly children, are undernourished.

Conway and Toenniessen (2003) discuss a case study of Mrs Namurunda. She is a single mother struggling to support a family in

Kenya. She farms a single hectare. The soils are well drained but acidic, highly weathered and leached of minerals. She needs one tonne of mixed crops just to survive and a further two to generate a modest income. Too often, however, she harvests less than one tonne per hectare (Figure 1a). Soil fertility is low in Africa as a result of millennia of soil erosion. Fertilisers are also expensive; US$400/tonne of urea versus US$90/tonne in Europe. In addition Mrs Namurunda's maize is attacked by the parasitic weed Striga, which sucks nutrition from its roots; by boring insects, which weaken the stems, and by the African indigenous Maize streak virus. Her cassava crop was devastated by cassava mealybugs and Cassava mosaic virus and her banana seedlings were infected with nematodes and the fungal disease black Sigatoka. Her beans were infected with fungal diseases and she faced a drought during the growing season that reduced the yield of all her plants.

Fortunately help is at hand. Agronomic improvements, biological control measures, and varieties improved by both conventional breeding and tissue culture techniques have led to the potential of Mrs Namurunda growing a secure crop (Figure 1b). Genetic engineering may be another factor that can improve food productivity and security in countries such as those on the African continent.

One argument put forward by the anti-GM crop lobby is that farmers will be forced to buy seed every season. This is a false argument (Thomson 2002). Many crops, including maize, depend on their yield due to hybrid seeds. These seeds are the product of two inbred parental lines, which, when crossed, produce hybrids of a high quality. If farmers plant seeds from these hybrids they will lose the quality of the parents. Thus farmers either choose to plant hybrids, in which case they will buy seeds every season, or they will plant their own seeds knowing that the yield will be less. There are a variety of 'open-pollinated varieties' of crops that have been specifically bred to allow farmers to plant their own seed. These have yields inferior to hybrids but are readily available to the farmer at a lower cost. However, the seeds can only be used for a few seasons before the yield is too poor to warrant planting. This process has been in place for decades, long before the advent of GM crops.

(a)

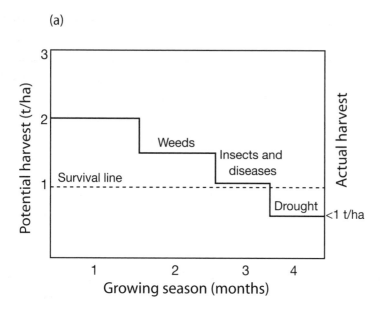

(b)

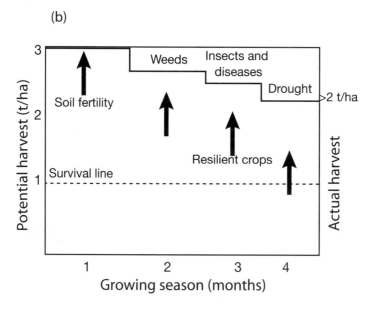

Figure 1 (a) An insecure farm in Africa. (b) A secure farm in Africa (Conway and Toenniessen 2003).

Currently, no open-pollinated varieties have been subjected to genetic modification. The reason is simple. There is very little economic incentive for seed companies to produce such seed. However, philanthropic organisations, such as the Rockefeller Foundation, are supporting research into GM open-pollinated varieties that would be reasonably available to small-scale farmers. At present, two-thirds of the maize seed sold globally is hybrid (James 2004).

Worldwide status of genetically modified crops

Genetically modified crops were commercialised in 1995–96 and the area under cultivation continues to increase (Figure 2). In 2005, ten years from the first commercialisation of GM crops, the increase over 2004 was 11%, equivalent to a total of 90 million hectares. The crops are grown in 21 countries, up from 17 in 2004. Three of the four new countries are in the European Union (EU), namely Portugal, France and the Czech Republic, which are now in addition to Spain and Germany.

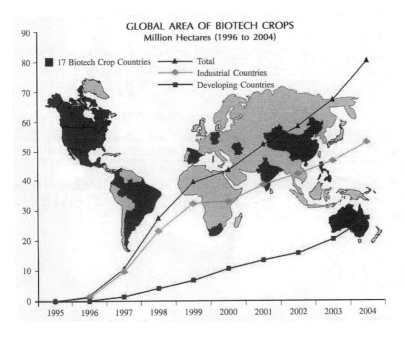

Figure 2 Global area of genetically modified crops (James 2005).

Iran was the fourth new country. Of the beneficiary farmers some 90% are in developing countries with limited access to resources. In 2005, the 21 countries included 11 developing and 10 industrial countries. The five leading developing countries are Argentina, Brazil, China, India and South Africa, planting more than one-third of the global area under cultivation with GM crops. The growth in plantings in these countries between 2004 and 2005 was 6.3 million hectares, representing a 23% growth. This was much higher than the corresponding growth in industrial countries, being 2.7 million hectares and 5% growth (James 2005).

This book attempts to answer the questions posed above regarding the possible environmental impacts of GM crops. Although it is clear to the reader that I support the development and use of GM crops, I hope that I have presented balanced arguments that will enable the reader to make up her or his mind.

References

Conway G and Toenniessen G (2003) Science for African food security. *Science* **299**, 1187–1189.

James C (2004) Global status of commercialized biotech/GM crops: 2004. ISAAA Brief no. 32. (International Service for the Acquisition of Agri-biotech Applications: Ithaca, NY.)

James C (2005) Global status of commercialized biotech/GM crops: 2004. ISAAA Brief no. 34. (International Service for the Acquisition of Agri-biotech Applications: Ithaca, NY.)

Thomson JA (2002) *Genes for Africa. Genetically Modified Crops in the Developing World*. pp. 45–46. (UCT Press: Cape Town, South Africa.)

1

Classical plant breeding and genetically modified crops

It is thought that about 7500 years ago farmers in Mexico domesticated a wild plant called teosinte. Over the centuries they selected varieties of this grass-like plant that had improved characteristics or traits, until today the offspring is maize, the crop planted to the second largest farmland area in the world. Teosinte and maize look so different that, until modern molecular genetic techniques were developed, the plants were classified in different genera. This process of selection between different, but related, plants is called plant breeding.

How does modern plant breeding work? Breeders take two different varieties of a plant, or plants from a variable population, each of which has individual characteristics or traits that make it attractive and cross-pollinate them. For example, one variety might produce high yields and the other might be resistant to an insect pest. The seed from the cross of the two varieties is collected and planted.

The plants that result contain a random sample of many thousands of **genes**, half originating from each of the parents. Plants that show both high yields and insect resistance are then selected by the breeders.

It is possible, however, that during this random process the selected plants might have acquired an undesirable trait that was not observed in either parent. This might be sensitivity to a **plant virus** that was not a problem in either parent but is found in the offspring. These plants would not be acceptable (Chrispeels and Sadava 2003, pp. 368–69).

The resulting plants are called **hybrids**, a combination of two sets of genes, one set originating from each parent. However, this is not usually the product the breeder is seeking. Usually the hybrid then needs to be **back-crossed** to one of the parent lines, called the recurrent parent because it is used repeatedly during crosses. The other parent is the donor because it contributes the desirable gene(s). In the example described here, the recurrent parent would be the high-yielding variety as this trait is multigenic. It is easier to introduce a trait that could be the product of a single gene, such as insect resistance, into a variety whose trait, such as high yield, results from the combined effects of many genes.

During **hybridisation**, the genes of the donor parent mix randomly with those of the recurrent parent. The purpose of back-crossing is to recover any desirable genes from the recurrent parent that may have been lost in the formation of the original hybrid plant. Each back-cross recovers an additional 50% of the genes of the recurrent parent, amounting to 75% after the first back-cross and 87% after the second. Breeders continue back-crossing until they have recovered the desirable level of genes from the recurrent parent. Usually it takes three back-crosses to recover virtually all the original genes (Thomson 2002).

Aids to plant breeding: mutations, jumping genes and horizontal gene transfer

As plant breeding is a slow process, breeders sometimes speed it up by introducing heritable changes into the **DNA** of one parent using **mutations**. Agents that cause mutations are called **mutagens** and can be chemicals or various types of radiation. As mutagens introduce random changes in the plant's DNA the results are far more unpredictable than changes brought about by specific genetic modification. For example, radiation in the form of gamma rays was used to alter the genes of a successful rice variety known as Calrose 76. The resulting plants are not radioactive. The radiation in this case, however, changed genes controlling plant height to reduce the height of the plants, resulting in increased yields of grain (Nuffield Council on Bioethics 2004).

A phenomenon that has assisted plant breeders is that DNA is inherently plastic. This means that DNA is not a static molecule but can undergo considerable natural rearrangements. These are caused by regions of DNA that can jump around an organism's **genome**. They are called **transposons** or jumping genes (McClintock 1951). Jumping genes can increase the variability and plasticity of an organism's genome. They provide breeders increased diversity with which to work. Random insertion of DNA into genomes occurs naturally in plants (Chrispeels and Sadava 2003, p. 129). For instance, if a piece of DNA jumps into genes involved in chlorophyll biosynthesis, the offspring of those cells could lack the normal green pigment, resulting in the variegation of foliage leaves found in some popular horticultural varieties.

Another natural phenomenon is the horizontal transfer of genes between species and even genera. **Vertical gene transfer** is what happens when genes are passed from parents to offspring. **Horizontal gene transfer** is the movement of genetic information between sexually unrelated organisms. As more and more genomes are sequenced it is apparent that horizontal gene transfer has occurred over the millennia. For instance, the bacterium that causes tuberculosis, *Mycobacterium tuberculosis*, carries a chunk of human DNA (Cole et al. 1998). Thus some genes have been transferred between different species in nature.

Genetic modification of plants

Although classical plant breeding has resulted in the development of a vast variety of crops, the process is largely hit-or-miss and rather slow. Genetic engineering allows scientists to transfer specific genes into plants resulting in the introduction of one or more defined traits into a particular genetic background. This process of **transformation** results in the production of a **transgenic plant**.

Although this process might be viewed as 'unnatural' compared with plant breeding, breeders often use other 'unnatural' methods, such as wide-crossing, which have led to completely new varieties such as triticale, a hybrid between wheat and rye, and the use of mutations as discussed above. The Nuffield Council on Bioethics (2004) report stated that 'it was not helpful to classify a crop that has been arrived at by means of conventional plant breeding as natural, and to classify a crop with the same genetic complement as unnatural if it has been produced through genetic modification'.

The report concedes that there is some concern that genetic modification might pose risks that differ from those of conventional plant breeding. For instance, unintended effects might occur when introducing a gene, for instance by disrupting existing genes. However, unintended effects are not specific to the use of genetic modification. They are often found in conventional breeding, especially in the case of mutation breeding.

One of the advantages of genetic modification is that genes can be taken from any organism. This can result in traits that would not normally be possible by breeding. Good examples of this are insect-resistant crops (see Chapter 2), herbicide-tolerant crops (see Chapter 3) and 'golden rice' enriched in pro-vitamin A (see Chapter 9).

Summary

A wide range of very successful crops has been produced by conventional plant breeding. However, it is somewhat hit-or-miss and slow. Genetic modification of plants speeds up the process by targeting the introduction of only specific genes. It also allows the use of genes from different organisms that, in some cases, introduces traits not normally possible in plant breeding.

References

Chrispeels MJ and Sadava DE (2003) *Plants, Genes, and Crop Biotechnology*. 2nd edition. (Jones and Bartlett: Sudbury, Massachusetts, USA.)

Cole ST et al. (1998) Deciphering the biology of *Mycobacterium tuberculosis* from the complete genome sequence. *Nature* **393**, 537–544.

Nuffield Council on Bioethics (2004) The use of genetically modified crops in developing countries. p. 23. (Nuffield Council on Bioethics: London, UK.)

McClintock B (1951) Chromosome organization and genic expression. *Cold Spring Harbor Symposium on Quantitative Biology* **16**, 13–47.

Thomson JA (2002) *Genes for Africa. Genetically Modified Crops in the Developing World.* (UCT Press: Cape Town, South Africa.)

2

Insect-resistant crops

Bacillus thuringiensis is a naturally occurring soil bacterium that produces proteins, called **Bt** proteins, that are toxic to certain insects (Schnepf et al. 1998). Over 40 varieties of these proteins have been identified. For instance some of the Bt toxins target the larval forms of lepidoterans (butterflies and moths), others dipterans (flies and mosquitoes) and others coleopterans (beetles). Bt proteins cause little or no harm to most non-target organisms, including people and wildlife. They have been used in sprays in conventional and organic agriculture for decades with little or no ill effects on the environment or human health. Thus, Bt toxins are considered environmentally friendly alternatives to broad-spectrum insecticides. However, these and other insecticide sprays are rather ineffective against pests that burrow into plant tissues and are not exposed to sprays. Beginning in the mid-1990s, crops expressing Bt genes and hence producing the toxins inside the plant, were commercialised in the United States (USA). The Bt toxins in the first generation of GM crops kill some of the key lepidopteran pests of maize (European corn borer, corn earworm, south-western corn borer) and cotton (cotton bollworm, tobacco budworm and pink bollworm).

Cotton

Economic benefits

Insect attack is one of the major constraints to cotton (*Gossypium hirsutum*), upland cotton, or *G. barbadense*, Sea Island or pima cotton) cultivation and the yield losses account for an estimated US$5 billion annually. About 25% of all insecticides used in agriculture are applied to cotton, more than to any other crop (James 2004). Furthermore this percentage can reach staggering proportions in some countries, such as central and west Africa (80%), Pakistan (79%) and India (48%). The reliance on insecticides in developing countries is high and in many cases represents a hardship for producers when the international price of cotton is low and when cotton is the principal cash crop (James 2002a). Moreover, cotton worldwide often receives more than one insecticide application per season – hence the potential for insecticide reduction in Bt cotton (ICAC 1995).

Bt cotton worldwide

In 2002 Bt-modified GM cotton was planted on 6.8 million hectares worldwide, comprising 12% of the world's production. In 2003 this figure increased to 7.2 million hectares, a 6% growth, resulting in Bt cotton accounting for 21% of the world's production. The growth in GM cotton plantings in China amounted to a 33% increase, GM cotton in the USA declined by 5% and that in Australia remained constant (James 2003). In 2004 the area planted to Bt cotton globally had increased by 1.8 million hectares, equivalent to a 25% growth over 2003 (James 2004). Most of the growth occurred in China, India and Australia. The take-up in China has been significant where about seven million resource-poor farmers, planting on average one-half to one hectare, have benefited from Bt cotton. In India the area approved for the crop was five times more in 2004 than in 2003.

Economic impacts of Bt cotton

There are two main ways in which the Bt trait can produce better economic returns for farmers. The first is by the reduction of losses of product, which will be reflected in increased yields. The second is by the reduction in the use of synthetic pesticide costs. Both are significant.

As an example, scientists in Germany and the USA have done extensive research into the economic impacts of Bt cotton in India where Bt cotton was only approved for commercialisation in 2003 . As a result the studies could not be market-based, but they did find that the yield of this crop can increase significantly, especially 'where pest damage is substantial and the use of alternative control agents is limited' (Qaim and Zilberman 2003). Bt cotton provides a fairly high degree of resistance to the American bollworm (*Helicoverpa armigera*), the major insect pest in India. They analysed data from field trials on 157 farms during the 2001/02 growing season. All farmers grew Bt cotton, the same variety but without the Bt gene and a local hybrid commonly grown in the particular district. The Bt varieties were sprayed three times less often against bollworm, reducing insecticide use by almost 70%. In addition almost all the pesticide reduction applied to the hazardous **organophosphates**, carbamates and **pyrethroids**. The difference in India was that the yields of Bt cotton were higher than the other two varieties. Average yields of the former exceeded the latter by 80 to 87%. This yield differential is higher than that found in China, Mexico or the United States. Qaim and Zilberman (2003) suggest that this is the result of high pest pressure in India and that farmers do not have access to affordable, effective insecticides.

Performance of Bt cotton in farmers' field trials in India showed a significant reduction in bollworm incidence compared with conventional cotton. In the Bt variety 11.5% of the fruiting bodies were damaged compared with 29.4% in the conventional one. Seven sprays of insecticide were made for control of insect pests on the conventional plants compared with three on the Bt ones. The results showed that Bt cotton provided higher returns although the initial seed cost for the farmers was higher (Bambawale et al. 2004).

In 2005, six new varieties of transgenic cotton were approved by the Indian Genetic Engineering Approval Committee for release, up from the three permitted for cultivation since 2003. These were grown on over 500 000 acres (about 202 500 hectares).

Scientists at the University of Reading in the United Kingdom (UK) have been weighing the economic costs and benefits of Bt cotton in South Africa for several years (Ismael et al. 2001; Morse et al. 2003). Seeds for this crop were released commercially in 1997 and have been grown since then rather extensively by small-scale farmers in the Makhathini Flats area of KwaZulu-Natal near the border with Mozambique. They estimated that by the 2001/02 growing season 90% of the farmers in this region were growing Bt cotton. Many of the main insecticides used here are highly toxic, and carrying a knapsack sprayer for many hours is tiring and dangerous to human health. Smallholder farmers, half of whom are women, receive a 77% higher return on Bt cotton (Table 2.1). In general, the smaller the farm, the greater the benefits received.

However, as the authors note, one does need to keep benefits in perspective. Bt cotton is not a magic bullet that will solve poverty among these farmers overnight. The highest gross advantage per hectare was found to be about ZAR700, which translates into about US$88 per hectare, depending on the rather volatile ZAR exchange rate. The agricultural wage rates in the Makhathini are about US$1 per day. Therefore, the advantage of growing Bt cotton is at best equivalent to about 15 to 18 days paid employment in a city – provided, of course, one is able to travel to the city and obtain work there. This is not an easy option in rural South Africa today. Therefore, if Bt cotton is managed properly to ensure continued productivity, it can be a positive and potentially sustainable contribution to farmers in this part of South Africa.

Table 2.1 Costs and benefits of GM cotton production among small-scale farmers in South Africa 1999–2000

	Conventional cotton	GM cotton
Yield (kg/ha)	261	417
Value (ZAR/ha)	568	905
Seed cost (ZAR/ha)	91	197
Pesticide cost (ZAR/ha)	116	72
Gross margin (ZAR/ha)	361	638

From Ismael et al. (2001); ZAR = US$6.52 during this period.

In a follow-up paper (Morse et al. 2004) comment that their analysis of the economic impacts of the uptake of Bt cotton by resource-poor, small-holder farmers in South Africa is the first such study on the continent of Africa based on farmers' own practice as distinct from field-trial data collected under controlled conditions. 'To our knowledge, there have been no comparable and large-scale studies on the continent of Africa and few anywhere that look at Bt cotton production under entirely farmer-managed conditions.'

They state that Bt cotton adopters achieved consistently higher yields and revenue per hectare than non-adopters over the three seasons 1998/99, 1999/2000 and 2000/01. This was particularly noticeable in the poor, wet growing season of 1999/2000, which favoured bollworm. Adopters had higher seed costs due to the Bt seed premium, but lower pesticide costs, both in product procurement and the use of labour for spraying. However, as a result of the higher yields achieved by the Bt adopters, their costs of harvesting, because of increased labour use, were significantly higher. This shows that the concerns of labour unions that the use of Bt cotton will result in the decrease of labour employed are unfounded under these conditions.

The result of all these differences between Bt cotton adopters and non-adopters is that the former achieved substantially higher gross margins across all three seasons. In financial terms this advantage amounted to about ZAR531 to 742 (about US$76–106) per hectare on average (Table 2.2). In the 1999/2000 wet season those growing non-Bt cotton actually had a negative gross margin, which resulted in them having difficulty in paying back loans (Morse et al. 2003).

Bt cotton adoption in Mexico varies widely and reflects regional patterns of pest infestation and the resultant economic losses (FAO 2004). Some areas are afflicted with pests that are not susceptible to Bt and thus require the use of chemical pesticides. Bt cotton adoption is therefore low in these regions. In addition, Bt cotton is barred from regions where wild species of *Gossypium hirsutum*, a native relative of cotton, exist (Traxler et al. 2003). In a study of the 1997 and 1998 growing seasons in the Comarca Lagunera region where Bt adoption is high the average yield difference between Bt and conventional cotton was

Table 2.2 Cotton yields, revenue and costs per hectare for adopters and non-adopters of Bt cotton in South Africa

Output (kg) and costs (SAR/ha)	1998/99		1999/2000		2000/01	
	Non-adopters (n = 1196)	Adopters (n = 87)	Non-adopters (n = 329)	Adopters (n = 112)	Non-adopters (n = 254)	Adopters (n = 254)
Yield	452	738	264	489	501	783
Total revenue	984	1 605	574	1 064	1 090	1 704
Seed cost	138	278	190	413	176	260
Pesticide cost	153	72	222	104	305	113
Spray labour cost	77	38	108	49	135	45
Harvest labour cost	113	184	66	122	125	196
Gross margin	502	1 033	−11	376	348	1 090
Gross margin gain in US$ based on exchange rate at harvest	86		59		93	

From Morse et al. (2004).

only 11%. Although pesticide costs were about 77% lower for Bt cotton, seed costs were almost three times higher. As a result the average profit differential for the two years was US$295 per hectare. The authors calculated the distribution of the economic benefits between the farmers and the seed companies and found that the farmers captured 86% of the total benefits.

Qaim and de Janvry (2003) studied the case of Bt cotton in Argentina during the 1999/2000 and 2000/01 growing seasons. Unlike herbicide-tolerant soybean (see Chapter 3), adoption of Bt cotton has been slow and by 2001 reached only about 5% of the total cotton-growing area. Part of the reason for this is that conventional varieties are better adapted to local conditions and have higher potential yields than the Bt varieties. In addition Bt seeds costs are more than six times greater than conventional varieties. Thus, net revenues for Bt varieties were higher by only a small margin. The performance differences between Bt and conventional cotton is shown in Table 2.3.

Table 2.3 Performance differences between Bt and conventional cotton

	Argentina	China	India	Mexico	South Africa
Lint yield (kg/ha)	531	523	699	165	237
%	33	19	80	11	65
Chemical sprays (No.)	−2.4	—	−3.0	−2.2	—
Gross revenue (US$/ha)	121	262	—	248	59
%	34	23	—	9	65
Pest control (US$/ha)	−18	−230	−30	−106	−26
%	−47	−67	—	−77	−58
Seed costs (US$/ha)	87	32	—	58	14
%	530	95	—	165	89
Total costs (US$/ha)	99	−208	—	−47	2
%	35	−16	—	−27	3
Profit (US$/ha)	23	470	—	295	65
%	31	340	—	12	299

From FAO (2004); all values are in US$.

Finally, on the question of economic impacts of Bt cotton, Brookes and Barfoot (2005) have analysed the results of the first nine years from 1996 until 2004 (Table 2.4).

Environmental impacts of Bt cotton

As stated earlier, cotton farmers worldwide are among the highest users of insecticides. The spraying of almost two million pounds of pesticides (roughly 50% of previous usage) has been spared since the large-scale adoption of Bt cotton. Bt cotton typically requires three to four sprays per season in the USA where about 80% of the cotton is Bt (James 2004). This is compared with five to 12 sprays for conventional cotton.

Table 2.4 Economic impacts of Bt cotton in the USA, China, Australia, Argentina, Mexico and India from 1996 until 2004

Country	Yield effect	Cost savings (/ha)
USA	9% 1996–2002 11% in 2003 and 2004	US$63.26 1996–2002 US$74.1 in 2003 and 2004
China	8% 1996–99 10% 2000 onwards	US$261 in 2000 US$438 in 2001 US$349.50 2002–04
Australia	None	A$151 in 1996 A$157 in 1997 A$188 in 1998 A$172 in 1999 A$267 2000–02 A$347 in 2003 and 2004
Argentina	30% all years	$17.47 all years
Mexico	Ranged from 3 to 37%	PS985 1996 and 1999–2004 US$121 in 1997 US$94 in 1998
India	45% 2002 63% 2003 54% 2004	Rs2,032 in 2002 Rs1,767 in 2003 Rs1,9000 in 2004

From Brookes and Barfoot (2005); all values are in local currency or US$.

Thus, there has been at best a reduction of nine, and at the least of one, spray per acre per season since the introduction of Bt cotton to the USA. This not only translates into less environmental pollution but also to cost savings as high as US$32 per acre in the cotton belt in Alabama (Marra et al. 2003).

Bt cotton was introduced into China in 1997. Previously farmers had been heavy users of insecticides, each spending on average US$100 per hectare per year. This translates into a massive US$500 million for the country (Huang et al. 2002). The banning of **organochlorines** in the early 1980s led to a switch to organophosphates and then to pyrethroids in the early 1990s; in both instances because the insects had developed resistance to the chemical pesticides. When resistance developed to pyrethroids China adopted cocktails of pesticides, which included DDT (Pray et al. 2002). By 2001 about 3.5 million farmers in

China were planting Bt cotton. This translated into 30% of the total cotton crop, and significantly, they were spraying 79 560 tonnes less of insecticide.

In 2002 Pray et al. published an analysis of the effects of Bt cotton in China during 1999, 2000 and 2001. They estimated that nearly 1.5 million hectares was planted in 2001 and as, on average, each farmer grew it on 0.42 ha, it is suggested that about 3.5 million farms had adopted Bt cotton in 2001. The mean yield of Bt cotton varieties was 5% to 6% higher than those of non-Bt cotton. In addition there appeared to be no obvious deterioration of the effectiveness of Bt varieties over time. Their data also demonstrated that Bt cotton varieties continued to reduce total pesticide use. As pesticide is primarily applied with small backpack sprayers, and as farmers usually do not use protective clothing (Figure 2.1), spraying can be a health hazard. Their data indicate that farmers planting non-Bt cotton reported a higher percentage of poisoning in each year (1999 and 2000, 22% and 29%) compared with non-Bt cotton (5% and 8% for the same years).

Figure 2.1 A Chinese farmer spraying his cotton field (courtesy of Carl Pray).

In a further analysis of Bt cotton in China, Wu and Guo (2005) concluded that planting these varieties has led to a 'drastic reduction in insecticide use, which usually results in a significant increase in populations of beneficial insects and thus contributes to the improvement of the natural control of some pests'. In addition there has been no trend towards the evolution of insect resistance to the Bt toxin despite intensive planting of Bt cotton (Wu and Guo 2005).

Insecticide use has also declined on smallholder farms in South Africa. Spraying has been cut from eight to two times per season reducing insecticide costs by one-third (Table 2.1; Ismael et al. 2001). A typical farmer, often a woman, is now spared 12 days of arduous spraying, saves over 1000 litres of water, and walks 100 km less per year to collect that water (James 2002a).

Moreover such a farmer is less likely to get sick. In the 1998/99 season local hospitals recorded 150 cases of burns and sickness due to agricultural chemicals. This dropped to 12 in 2000/01 when Bt cotton was widely grown in the region (Thirtle et al. 2003).

In a six-year field study Naranjo (2005) assessed the long-term effects of Bt cotton on the abundance of non-target insects, including natural enemies of cotton bollworm. They found no effects of Bt cotton in contrast to a much greater negative effect due to insecticide applications. They concluded that 'the effects of Bt cotton on a representative non-target community are minor, especially in comparison with the alternative use of broad-spectrum insecticides'.

On this aspect of Bt cotton, in their analysis of their overall environmental impacts from 1994 until 2004, Brookes and Barfoot (2005) concluded that major environmental gains have been derived from these plantings. Indeed, these gains have been the largest of any GM crops on a per hectare basis. 'Since 1996, farmers have used 77 million kg less insecticide on Bt cotton which translates to a 15% reduction.'

Finally, in a recent two-year farm-scale evaluation of the impacts of Bt cotton on pesticide use in 81 commercial fields in Arizona, Cattaneo et al. (2006) found that transgenic cotton was treated with fewer broad-spectrum insecticides than non-transgenic cotton.

Is Bt cotton safe?

In 2001 the US Environmental Protection Agency (EPA 2001) reassessed the registration of four Bt crops whose registration was soon to expire. The reassessments included provisions to prevent gene flow from Bt cotton to weedy relatives, increased research data on potential environmental effects and strengthening insect-resistant management (Mendelsohn et al. 2003). The EPA decided that Bt cotton and Bt corn do not pose unreasonable risks to human health or to the environment (EPA 2001).

Before coming to this conclusion, the EPA considered the data on the development of resistance to the Bt toxin among target insects, gene flow from Bt plants and the potential for weeds to develop if pollen from Bt crops were to fertilise other plants (see also Chapter 6), horizontal gene transfer (see also Chapter 7), the effect of pollen from Bt plants on non-targeted insects such as butterflies, and the fate of Bt proteins in the environment.

Insect resistance to Bt cotton

The unrestricted use of Bt crops such as cotton is likely to lead to the evolution of resistance in the target insect pests unless measures are taken to delay or halt its development. The loss of Bt as an effective pest management tool could have adverse consequences for the environment as growers might shift toward using more toxic pesticides. In addition organic farmers might lose a valuable biodegradable pesticide in the form of the Bt bacteria themselves.

Scientists and skeptics of Bt crops alike expected that target insect pests would evolve resistance to these crops. Defying these expectations, scientists funded by the US Department of Agriculture (USDA) have found surprisingly little resistance in the field. This suggests that transgenic Bt crops could have a longer and more profitable commercial life than expected. Moreover, it would appear that the practices that have been adopted to diminish levels of insect resistance to the Bt toxin are working. As quoted in Fox (2003) entomologist Bruce Tabashnik of the University of Arizona in Tucson, USA, said, 'If I'd gotten up seven years ago and said that there would be no evidence of increased Bt resistance after Bt crops were planted on 62 million hectares [cumulative and worldwide], I would have been hooted off the stage'. He further states that although more than 500 species of insect have evolved resistance to one or more conventional insecticides, only one pest, the diamondback moth, has evolved resistance to Bt sprays, and none to Bt crops. Although the success of Bt crops has exceded the expectations of many,it does not mean that resistance could not develop in the future (Tabashnik et al. 2003).

The primary measure taken to limit resistance development in insects is that farmers planting Bt crops are required to set aside a certain acreage that they plant with non-Bt varieties. These varieties act as **refuges** that permit susceptible insects to survive and swamp out resistant variants that might emerge from insects feeding on the neighbouring Bt plants. The susceptible insects can multiply on the refuges, compete with and mate with resistant insects, effectively diluting out the resistance genes in the next generation. However, this strategy works best if the transgenic plants produce enough toxin to kill hybrid progeny produced by matings between resistant and susceptible insects. If the crop produces too little Bt toxin, the survivors will include many hybrid, partially resistant insects likely to mate with each other and enrich the gene pool for resistance. Current varieties of Bt cotton produce high enough levels to kill such hybrid progeny of some pests, but not others (Tabashnik et al. 2003).

Tabashnik et al. (2005) confirmed these results. They monitored pink bollworm (*Pectinophora gossypiella*) resistance to Bt toxin for eight years with laboratory bioassays of strains derived annually from 10 to 17 cotton fields in Arizona. They chose this insect as it is a major cotton pest in this state and has experienced intense selection for resistance to Bt cotton since 1997. Their results showed no net increase from 1997 to 2004 in the frequency of pink bollworm resistance to Bt toxin.

Despite the seeming effictivity of the use of refuges, scientists are still on the look out for additional ways to limit resistance development. One such measure could be the development of crops expressing not one, but two different Bt toxins. These would target the same insect pest but act in slightly different ways. Thus insects would have to develop simultaneous resistance to both toxins, a much rarer event than the development of resistance to a single toxin. Scientists tested this idea on broccoli plants under laboratory conditions where insects are able to develop resistance. They found that dual toxins significantly slowed down the emergence of insect resistance (Zhao et al. 2003). Such 'double-stacked' cotton plants have been introduced into Australia with considerable success (D. Tribe, pers. comm. 2005).

A more recent approach has been to fuse the Bt toxin gene with the region encoding the non-toxic domain of the toxic ricin protein found in castor beans. The resulting protein binds to carbohydrate residues in the lining of insect guts and enhances the binding of the Bt toxin. This increases the effectiveness of the Bt toxin and also decreases the likelihood of insects evolving resistance as mutations would have to occur in the genes coding for both binding receptors simultaneously. Results have shown that transgenic rice and maize plants expressing this fusion protein are significantly more toxic to a range of insect pests than those expressing the Bt toxin alone. In addition, the fusion protein conferred resistance to a broader spectrum of insect pests (Mehlo et al. 2005).

Gene flow and weediness

Bt plants that have been registered to date do not have a reasonable possibility of passing this trait to wild native plants. This is because of differences in chromosomes, the timing of pollination and habitat. The exception is the possibility of gene transfer from Bt cotton to wild cotton relatives in Hawaii, Florida, Puerto Rico and the US Virgin Islands. The EPA has restricted the sale or distribution of Bt cotton in these areas (Mendelsohn et al. 2003).

Horizontal and vertical gene transfer

Vertical gene transfer is a term normally understood to mean the transfer of genes from parent to offspring of the same species. Horizontal gene transfer is a term that is used for the transfer of genes from one organism to another, often unrelated, species. For example, can genes be transferred from a Bt crop, such as cotton, to a soil microorganism and if so, what are the risk implications? Many experiments have been published which show that this is extremely unlikely under field conditions (Nielsen and Townsend 2004). Indeed horizontal gene transfer has only been detected under conditions designed to specifically favour such transfer (see Chapter 7). In addition genes such as the one encoding Bt have their origins in soil bacteria. Thus, the EPA concluded that horizontal gene transfer is at most an extremely rare event and that the Bt trait is already present in soil bacteria (Mendelsohn et al. 2003).

Exposure of soil organisms to Bt toxins
There are several routes whereby soil organisms can be exposed to Bt proteins. These include leakage from roots and tilled plant tissues, and feeding on living or dead roots and other plant organs. Although most of the Bt protein deposited in soil is degraded within a few days, some soil components such as clays appear to bind these proteins, thus preventing their degradation by soil microorganisms (Stotzky 2002). Nonetheless, available data indicate that Bt proteins do not have any measurable effects on the microbial populations in the soil. These include bacteria, fungi, protozoa, algae, springtails and earthworms. In addition, plants planted in soil containing Bt proteins do not take up the toxin (Mendelsohn et al. 2003).

Maize

Maize (*Zea mays*) is one of the world's earliest agricultural innovations. It was probably domesticated in Mexico some 10 000 years ago when farmers selected plants with superior characteristics. Slowly they turned a scraggly, non-descript grass called teosinte into modern productive maize. The first GM Bt maize was planted in the USA in 1996 and accounted for 0.3 million hectares. In 2003 it was planted on 15.5 million hectares worldwide and was responsible for the highest percentage growth of any transgenic crop. The increase over 2002 amounted to 25% compared to a 20% increase for canola, 13% for soybean and 6% for cotton (James 2003). The 25% increase was repeated in 2004, much of it in the USA, but also in Canada, Argentina, Spain and South Africa (James 2004).

One of the problems of growing maize in developing countries is the poor yield, particularly in Africa. Whereas in the USA the average yield in metric tonnes per hectare was 8 in 2002, the yields in Africa range from 0.8 (Zimbabwe and the Democratic Republic of Congo) to 2.7 (South Africa) with an average of 1.4 in the nine countries recorded (James 2002). GM maize may help to increase these yields in developing countries.

There are several types of maize planted by farmers around the world. These range from local, traditional **landraces** used by subsistence farmers, to selected **open-pollinated varieties (OPV)** used by more progressive farmers, to hybrids used by the most advanced farmers in developing countries and considered the norm in industrial countries (James 2002b). Farmers using the first type will plant their own seed saved from season to season, those using the OPVs can plant their own seed for a few seasons before they lose their competitive advantage, whereupon they will buy new seed. Farmers planting hybrid maize will buy their seed every season. The percentages of areas planted to these three types in selected countries are shown in Table 2.5.

Table 2.5 Areas (%) sown to maize hybrids, open-pollinated varieties (OPVs) and farmer-saved seed in selected countries in 1999

Region	Hybrids	OPVs	Farmer-saved seed
East and South Africa	81	11	8
North Africa	9	38	54
South East Asia	35	36	29
China	84	6	10
Mexico and Central America	15	8	77
South America	62	12	26
West Europe	98	2	0
USA/Canada	100	0	0

From James (2002b).

On a global basis 80% of the maize area of 140 million hectares is sown to improved varieties (hybrids and OPVs) with about two-thirds sown to hybrids. Many farmers in developing countries use improved varieties and can therefore have access to GM improved maize varieties should they wish to plant these.

Economic impact of Bt maize

Brookes and Barfoot (2005) undertook an analysis of the economic impacts of Bt maize in several countries from its initial plantings in 1996 until 2004 (Table 2.6).

Table 2.6 Economic impacts of Bt maize in the USA, Canada, Argentina, the Philippines and Spain from 1996 until 2004

Country	Yield effect	Cost savings
USA	5% all years	US$15.5 all years
Canada	5% all years	US$15.5 all years
Argentina	9% all years	Nil all years
Philippines	25% all years	PS800 in 2003 and 2004
Spain	6.3% all years	42 Euros all years

After Brookes and Barfoot (2005).

Environmental benefits of Bt maize

The first generation of GM crops were developed to help farmers decrease their expenditure not only by increasing yields as a result of reducing the impact of pests, but also by spending less on insecticides. The latter, however, is not as important a factor with maize as it is with cotton. Insecticides are not very effective in controlling maize borers that conceal themselves in the stems and cobs where insecticides cannot penetrate. Hence most maize farmers do not use insecticides, despite crop losses resulting from borers. In addition, seasonal variations in the numbers of borers are considerable. Thus before the introduction of Bt maize in 1996, only 2% of maize in the USA was sprayed.

Is Bt maize safe for the environment?

Gene flow

The risks associated with gene flow from GM crops, such as creating more invasive or new weeds, reducing **biodiversity**, or harming non-target species, are similar to those of conventionally bred crops. There are few instances of crop plants becoming weeds. This is because crops have been bred so intensively for hundreds of years that they cannot survive without human intervention. However, increased weediness could occur if a GM plant were to become fitter than its conventional counterpart and thereby able to out-compete other crop species. They could do this by producing more seed, by dispersing pollen or seed further, or by growing more vigorously in a specific environment. Wild varieties of modern crops do not often coexist where the crop is grown. However, for maize this is not true as cultivated varieties are grown close to the local varieties, or landraces, in Mexico. There is a fear that transgenic varieties with competitive advantage might gradually displace the valuable genetic diversity inherent in such landraces.

The term that is used for the transfer of genes from one variety of a crop to another is **introgression**. This occurs mainly with naturally **cross-pollinating** crops, such as maize, as opposed to those that are self-fertilising, such as most other cereals except rye. It is generally accepted that unless specific measures are taken to prevent introgression, it will occur regardless of whether or not the maize is GM. An example of the prevention of introgression is the measure taken by seed maize farmers who have to be able to certify that the maize seed they produce has not been contaminated by different maize varieties grown nearby, either on their own or neighbouring farms. Seed maize farmers in the USA have found that, based on long-term experience, a 220-metre separation is sufficient to prevent such introgression (Jarvis and Hodgkin 1999). But what about introgression into landraces in Mexico?

It is important to understand how landraces and other maize varieties are grown and managed by farmers in Mexico. Here farmers grow different landraces close to each other under conditions where outcrossing is high. It is estimated that one-third of local maize varieties have introgressed genes from non-local or improved varieties (Gonzalez and

Goodman 1997). Thus, the view that landraces are pristine sources of biodiversity is erroneous and the possible introgression of transgenes should be seen within the broader context of the massive genetic exchange that has already occurred in Mexican maize landraces over a long time. Mexican farmers are not passive bystanders in this process. Many of them take an active role in facilitating the introgression of genes from improved varieties into their local landraces. It is highly unlikely that any landrace in production today is the same as its counterpart even 20 years ago.

However, recognising that there is a need to conserve the genetic variability of landraces, gene banks have been established to conserve seeds for use in current and future breeding programs. The International Centre for Wheat and Maize Improvement (CIMMYT), based in Mexico, has one such large bank.

Insect resistance to Bt maize

Until late in 2002 the only commercially available strategy for minimising the chance that targeted insects would become resistant to Bt crops was the use of refuges (see Insect resistance to Bt cotton). However, this relies on farmers planting the correct percentage of non-Bt crops in the field. On-farm assessments and telephone surveys in the USA in 2005 indicted that over 90% of growers adhered to the refugee requirement (Sayler 2006). Thus this may not be a problem in developed countries, but how will regulatory bodies manage this in developing countries? An alternative method approved for use in Australia and the USA in 2002 involves the stacking or pyramiding of two different Bt genes. Several studies show that this approach significantly delays the onset of insect resistance (Zhao et al. 2003).

Effect on non-target wildlife

The 'storm in a teacup' case of Bt maize harming the Monarch butterfly in the USA has been discussed by Thomson (2002). The US EPA reviewed the data on potentially harmful effects on other wildlife species. The data (Table 2.7) show no differences in the number of total insects or the numbers of insects of different insect types between the transgenic and conventional crop plots. No shifts in distribution of insects were seen except where the predator insects depended on the relevant Bt-targeted pest insect as its major food source. These tests were conducted in the absence of insecticide spraying, during which both prey and predator would be targeted (Mendelsohn et al. 2003). Indeed in studies where some of the conventional maize fields have been sprayed for the target pest of Bt maize, many non-target species have been killed leading to significantly lower non-target populations, at least transiently, in sprayed conventional fields as compared to unsprayed Bt maize fields.

Romeis et al. (2006) have studied the effects of a variety of Bt crops on non-target insects, including those used as biological control organisms. Field studies confirmed that the abundance and activity of these insects are similar in Bt and non-Bt crops. In contrast, applications of insecticides usually resulted in negative impacts on biological control organisms. They conclude that 'because Bt transgenic varieties can lead to substantial reductions in insecticide use in some crops, they can contribute to integrated pest management systems with a strong biological control component'.

Table 2.7 No adverse effect of Bt maize crops were found on the following non-target wildlife

Test materials	Non-target wildlife
Cornmeal	Quail
Corn pollen	*Daphnia magna* (water flea)
Corn pollen	*Daphnia magna*
Bt protein	Honey bee adults and larvae
Corn pollen	Honey bee larvae
Bt protein	Ladybird beetle
Bt protein	Parasitic wasps
Bt protein	Green lacewing
Bt protein in soil	Collembola (spring tails)
Bt protein in soil	Earthworms

After Mendelsohn et al. (2003).

In a more recent study on earthworms (Vercesi et al. 2006), no delete-rious effects were found when Bt corn leaves were added to soil. This was also true when worms were kept in pots with a growing Bt-corn plant.

Human health benefits of Bt maize

Most maize consumed by humans in the developed world is either milled, processed or purchased as corn-on-the cob. In the former two cases the products are subjected to food safety tests, and in the latter mere inspec-tion by the purchaser is sufficient to ensure quality. However, this is not the case with consumers in the developing world. Take a subsistence maize farmer in Africa. She or he harvests the crop and stores it year round for consumption by the family and possibly close neighbours. The maize is usually stored in roofless structures where it is subjected to sunshine and rain. If the maize has been attacked by insects that bore into the ears, the holes, together with the moist, warm conditions, provide the ideal breeding ground for fungi. These will cause rotting of the ears, but more importantly, they can produce life-threatening toxins. These **myco-toxins** (myco = fungi) include **fumonisins**, produced by *Fusarium* spp. and **aflotoxins**, produced by *Aspergillus* spp., among others (Marasas 2001). The effects these mycotoxins have on humans include oesophageal cancer, liver cancer and neural tube defects (Hendricks 1999; Wang et al. 2003). Populations at special risk appear to be in southern Africa (Marasas 1996), the Linxian region of China (Li et al. 1980) and north-eastern Italy (Franceschi et al. 1995).

Can Bt maize help? It is possible due to the fact that these kernels have decreased levels of insect damage. Most of the studies have been carried out in Europe where it was found that over several years Bt maize consistently decreased the level of fumonisins (Munkvold et al. 1999). On average there was more than a six-fold decrease in levels (see James 2002b). Bt maize allowed the levels of mycotoxins to be lowered to below the acceptable levels of 2 ppm compared with the levels of 9 ppm in conventional maize. The latter is almost five times higher than the accept-able guidance levels specified by the World Health Organization. The effect of Bt maize on the prevalence of mycotoxins in stored maize in Africa and other developing countries awaits further study, but it would appear that this crop could enhance the quality and safety of maize for animal and human consumption (Marasas and Vismer 2003).

The impact of Bt maize in South Africa

South Africa is the only developing country in which a basic subsistence crop – white maize – has been genetically modified. Bt yellow maize, primarily used for cattle feed and in the food industry, has been on the market since 1998, but Bt white maize has only been available commercially since 2001. This GM maize could be an important case study to determine whether GM crops can benefit commercial farmers as well as small-scale farmers and poor consumers.

The early stages of adoption of GM maize were slow. By the 2002/03 season 20% of yellow maize but only 2.8% of white maize was Bt. There were three major constraints that slowed the spread of Bt maize. The first was that the Bt hybrids on the market were not the best for African markets or agricultural conditions. The second was that many farmers did not see a large productivity advantage from planting Bt maize as planting strategies can mitigate against the stalk borer in areas other than those most affected by the pest. The third reason was farmers' concerns that they would not be able to sell their crop due to consumer concerns about GM food.

This started to change in 2000 and 2001 with the introduction of yellow maize hybrids specifically developed for South Africa's dry windy conditions. In addition, in 2001/02 borers were a serious problem for commercial farmers. This led to increased demand for Bt maize in 2002/03. And finally farmers did not have any difficulty in selling their GM crops. As a result the demand for GM maize has increased (Table 2.8). Indeed the major constraint may be the supply of seed, especially white maize, which is not keeping up with the demand. The yields of GM crops from 2000 to 2005 are shown in Table 2.9.

Planting of GM maize increased its market share from 14.6% in 2005 to 29.4% in 2006. White maize increased from 8.6% of total area planted in 2005 to 28.8% in 2006 (Van der Walt 2006).

Most of the plantings of Bt yellow and white maize in South Africa are by large-scale farmers. Small-scale farmers who plant maize for home consumption plant only white maize as this is traditionally consumed by black South Africans. A few farmers plant Bt yellow maize to feed to animals. Although a few small-scale farmers have planted Bt white maize, commercial farmers bought almost all the limited supply of this seed in 2002/03.

Table 2.8 Estimated area planted to transgenic maize in hectares and percentage of the total crop in South Africa

	1999/2000	2000/01	2001/02	2002/03
Yellow maize	50 000	75 000	160 000	197 000
% Total	3	5	14	20
White maize	0	0	6 000	55 000
% Total	0	0	0.4	2.8
Total Bt maize	50 000	75 000	166 000	252 000
% Total	1.3	2.3	4.7	7.1

From Gouse et al. (2005).

Bt seeds are sold at a premium of ZAR230 per 60 000 seeds (US$35 at an exchange rate – which can be quite volatile – of about 6.5) more than conventional hybrids. The top-of-the-line conventional hybrids sell at ZAR850 to 900 meaning that commercial farmers buying these lines will pay a surcharge of 20% to 21%. Lowest quality maize sells at ZAR650 per 60 000 seeds, meaning that small-scale farmers buying Bt maize will pay a surcharge of 26%. They will therefore have to show considerable profit to continue to plant these seeds. Time will tell if this is the case.

Commercial farmers who have planted Bt yellow maize have seen an increase in their income compared with their conventional maize fields, despite paying the technology fee. As they received the same

Table 2.9 Yields of GM crops (1000 hectares) in South Africa from 2000 to 2005

Crop	2000	2001	2002	2003	2004	2005 (estimated)
Bt maize (white)	Not planted	4	61	155	233	308
Bt maize (yellow)	77	125	109	170	210	205
Bt cotton	18	25	24	39	25	30
Herbicide-resistant soybeans	Not planted	Not planted	13	45	50	60
TOTAL	95	154	207	409	520	603

From W Green (pers. comm. 2006)

price for their Bt maize and their conventional maize, the difference was directly due to the yield increase. They were also able to reduce their costs by reducing the amount of pesticide applied. The increase in net income ranged from US$24 per hectare in dryland areas to US$143 per hectare in irrigated regions (Gouse et al. 2005).

Can small-scale farmers benefit from Bt white maize? With price differentials of US$83/kg Bt maize seed compared with $52/kg for non-Bt seed the answer is probably no, unless the farmers already buy hybrid maize. It is estimated that only 10% of small-scale farmers use hybrid seeds, the rest planting open-pollinated varieties and saved seeds of open-pollinated varieties. These are varieties bred specifically to allow farmers to replant seeds for a limited number of seasons before buying fresh seed again. None of these varieties have been genetically modified, largely because the profit margins are very low. However, there may be more opportunities for the adoption of Bt white maize by small-scale farmers elsewhere in Africa where many have already adopted hybrids. These countries include Zimbabwe where 91% of the maize area was hybrid in 1997–99, Kenya with 85% and Zambia with 65% (Pingali 2001). That said, when small-scale farmers in South Africa do plant hybrid maize, and they are now planting both Bt and herbicide-resistant maize, they would prefer to plant one variety carrying both traits (Figure 2.2).

A survey was conducted among 368 small-scale farmers in six sites in South Africa during the 2001/02 growing season (Gouse et al. 2005). The evidence suggests that Bt maize has potential benefits for such farmers (Figure 2.3). Another important finding was that the farmers also liked the quality of the maize produced by the Bt variety, Yieldgard®. At harvest, farmers were shown their own seed, seed from the same variety as the Bt maize but without the Bt gene and Yieldgard®. They were asked to judge the grain according to quality. Most farmers rated Yieldgard® grain as excellent whereas many farmers rated the others as good. Farmers at three sites chose quality as the best feature, whereas farmers at three other sites chose yield. Not much importance was given to pesticide reduction, probably because only half of them used pesticides (Gouse et al. 2005).

Figure 2.2 A South African small-scale farmer holding insect-resistant maize in his left hand and herbicide-resistant maize in his right hand (courtesy of David Tribe).

Can Bt maize spread to a larger share of small-scale farmers in South Africa? One solution would be for seed companies to charge a lower technology fee to these farmers compared to commercial farmers. One company is already doing this with conventional hybrid maize seed. Another way would be for private–public partnerships to introduce the Bt gene into open-pollinated varieties. This would enable small-scale farmers to save their seed and still get the benefit of Bt. However, it would be impossible for the government to enforce any type of Bt maize refuge plantings. As such plantings delay the onset of the evolution of Bt-resistant stem borers, this could increase the speed at which resistant insects develop. Thus, until we know more about the development of resistance, this possibility is probably not a realistic option (Gouse et al. 2005).

It is important that the biosafety regulatory process does not make it more expensive to provide Bt maize to small-scale farmers. Under the current system in South Africa the companies that sell GM seeds must

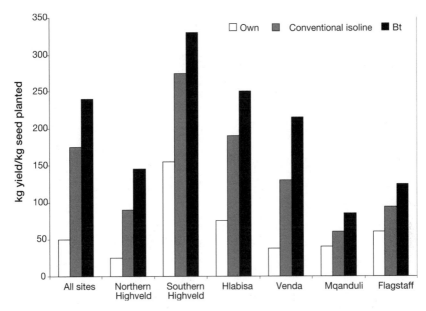

Figure 2.3 Small-scale farmers' yields per kilogram of seed, by seed type for six sites in 2001/02 (Gouse et al. 2005).

sign a contract with every farmer to ensure s/he plants the seeds only in the designated area and abides by the proper refuge requirements. They are required to plant 20% of their Bt area with non-Bt hybrids if they spray the refuges with pesticides, or 5% if they do not. This is relatively easy for large companies who sell directly through their marketing agents to large commercial farmers. However, this is an expensive requirement when companies are dealing with thousands of small-scale farmers and could well lead to a decision not to sell GM seeds to them at all.

To bring insect-resistant maize to the rest of Africa, a consortium is undertaking a project entitled IRMA (Insect Resistant Maize for Africa). This aims to provide sub-Saharan Africa small-holder maize producers with access to suitable Bt maize varieties that are resistant to the major stem borers that limit maize productivity in the region. A combination of traditional plant breeding and genetic modification is being used in this project.

Bt crops in the pipeline

Rice is a major source of energy and protein in Asia. Pests include stem borers but breeders have not produced resistant varieties. Several laboratories have produced Bt lines but none has been commercially released to farmers (Shelton et al. 2002). However, in China farm-level pre-production trials are underway. The effects of two locally produced varieties have been studied (Huang et al. 2005). One carries the Bt gene and the other an inhibitor of trypsin, an enzyme in the insect gut that helps digest proteins. Table 2.10 shows the pesticide use and yields of rice between farmers adopting the GM varieties (adopters) and those that did not (non-adopters).

Table 2.10 Pesticide use and yields of insect-resistant GM rice adopters and non-adopters in preproduction trials in China 2002–03

Parameter	Adopters	Non-adopters
Pesticide sprays (no. of times)	0.5	3.7
Expenditure on pesticides (yuan/ha)	31	243
Pesticide use (kg/ha)	2.0	21.2
Pesticide spray labour (days/ha)	0.7	9.1
Rice yield (kg/ha)	6364	6151
Number of observations (plots)	123	224

From Huang et al. (2005).

In addition households were asked how the use of pesticides affected their health. None of the adopters reported that their health was adversely affected while 8.3% of non-adopters in 2002 and 3% in 2003 reported adverse effects. This study provides evidence that there are positive impacts of GM rice on productivity, pesticide use and farmer health.

Apart from rice borers, the corn rootworm complex (*Diabrotica*) is also being targeted. Other Bt crops under development include canola, tobacco, tomatoes, apples, soybeans, peanuts, broccoli, cabbage and potatoes. Clearly there is a need for such insect-resistant crops.

Summary

Bt cotton plantings have increased in China, India and Australia, particularly among resource-poor farmers. This has resulted in decreases in the amount of insecticides used. Insect resistance to Bt has not yet become a problem but this is possibly only a matter of time. This emphasises the importance of resistance management with the use of refuges. Bt maize is also growing in popularity in the USA, Canada, Argentina, Spain and South Africa. Gene flow into landraces and other maize varieties has been reported. It is possible that Bt maize might help to decrease the levels of mycotoxins in maize grown by subsistence farmers. Unlike in the case of Bt cotton, it is likely that commercial farmers are most likely to benefit from Bt maize.

References

Bambawale OM, Singh A, Sharma OP, Bhosle BB, Lavekar RC, Dhandapani A, Kanwar V, Tanwar RK, Rathod KS, Patange NR et al. (2004) Performance of *Bt* cotton (MECH-162) under integrated pest management in farmers' participatory field trial in Nanded district, Central India. *Current Science* **86**, 1628–1633.

Brookes G and Barfoot P (2005) GM Crops, the global economic and environmental impact – the first nine years 1996–2004. *AgBioForum* **8**, 187–196.

Cattaneo MG, Yafuso C, Schmidt C, Huang C-y, Rahman M, Olson C, Ellers-Kirk C, Orr BJ, Marsh SE, Antilla L, Dutilleul and Carrière Y (2006) Farm-scale evaluation of the impacts of transgenic cotton on biodiversity, pesticide use, and yield. *Proceedings of the National Academies of Sciences of the USA* **103**, 7571–7576.

EPA (2001). Pesticides, Regulating Pesticides. (USEPA). http://www.epa.gov/pesticides/biopesticides/index.htm [Verified 19 April 2006].

FAO (2004) *The State of Food and Agriculture*. (Food and Agriculture Organization of the United Nations: Rome.)

Fox JL (2003) Resistance to Bt toxin surprisingly absent from pests. *Nature Biotechnology* **21**, 958–959.

Franceschi S, Bidoli E, Baron AE and La Vacchia C (1995) Maize and risk of cancer of the oral cavity, pharynx and esophagus in northeastern Italy. *Journal of the National Cancer Institute* **82**, 1407–1411.

Gonzalez FC and Goodman, MM (1997) Research on gene flow between improved maize and landraces. In *CIMMYT. Gene Flow among Maize Landraces, Improved Maize Varieties, and Teosinte, Implications for Transgenic Maize.* pp. 67–72. (CIMMYT: Mexico.) http://www.cimmyt.org/ABC/geneflow/geneflow_pdf_engl/ Geneflow_ResGene.pdf [Verified 5 June 2006].

Gouse M, Pray CE, Kirsten J and Schimmelpfennig D (2005) A GM subsistence crop in Africa, the case of Bt white maize in South Africa. *International Journal of Biotechnology* **7**, 84–94.

Hendricks K (1999) Fumonisins and neural tube defects in South Texas. *Epidemiology* **10**, 198–200.

Huang J, Rozelle S, Pray C and Wang Q (2002) Plant biotechnology in China. *Science* **295**, 674–677.

Huang J, Hu R, Rozelle S and Pray C (2005) Insect-resistant GM rice in farmers' fields, assessing productivity and health effects in China. *Science* **308**, 688–690.

ICAC (1995). Agrochemicals used on cotton. International Cotton Advisory Committee, Washington, DC. http://www.icac.org [Verified 18 April 2006].

Ismael Y, Bennet R and Morse S (2001) Farm level impact of Bt cotton in South Africa. *Biotechnology and Development Monitor* **48**, 15–19.

James C (2002a) Global review of commercialized transgenic crops, 2001 Feature, Bt cotton. ISAAA Brief no. 26. (International Service for the Acquisition of Agri-biotech Applications: Ithaca, NY.)

James C (2002b) Global review of commercialized transgenic crops, 2002 Feature, Bt maize. ISAAA Brief no. 29. (International Service for the Acquisition of Agri-biotech Applications: Ithaca, NY.)

James C (2003) Global status of commercialized transgenic crops, 2003. ISAAA Brief no. 30. (International Service for the Acquisition of Agri-biotech Applications: Ithaca, NY.)

James C (2004) Global status of commercialized Biotech/GM crops, 2004. ISAAA Brief no. 32. (International Service for the Acquisition of Agri-biotech Applications: Ithaca, NY.)

Jarvis DI and Hodgkin T (1999) Wild relatives and crop cultivars. Detecting natural introgression and farmer selection of new genetic combinations in agroecosystems. *Molecular Ecology* **8**, S159–S173.

Li M, Lu S, Jin C, Wang Y, Wang M, Cheng S and Tian G (1980) Experimental studies on the carcinogenicity of fungus-contaminated food from Linxian County. In *Genetic and Environmental Factors in Experimental Human Cancer*. (Ed. HV Gelboin.) pp. 139–148. (Japan Science Society Press: Tokyo.)

Marra M, Pardey P and Alston J (2003) The payoffs to transgenic field crops, an assessment of the evidence. *Agbioforum* **5**, 43–50. http://www.agbioforum.org [Verified 18 April 2006].

Marasas WFO (1996) Fumonisins, history, world-wide occurrence and impact. In *Fumonisins in Food*. (Eds LS Jackson, JW DeVries and LB Bullerman.) pp.1–17. (Plenum Press, New York.)

Marasas WFO (2001) Discovery and occurrence of the fumonisins. *Environmental Health Perspectives*, **109**(2), 239–243.

Marasas WFO and Vismer HF (2003) Advances in stored product protection. *Proceedings of the 8th International Working Conference on Stored Product Protection*. 22–26 July 2002. (Eds PF Credland, DM Armitage, CH Bell, PM Cogan and E Highley.) pp. 423–427. (CAB International, Wallingford, UK.)

Mehlo, L, Gahakwa D, Nghia PT, Loc NT, Capell T, Gatehouse JA, Gatehouse AMR and Christou P (2005) An alternative strategy for sustainable pest resistance in genetically enhanced crops. *Proceedings of the National Academies of Sciences of the USA* **102**, 7812–7816.

Mendelsohn M, Kough J, Vaituzis Z and Matthews K (2003) Are Bt crops safe? *Nature Biotechnology* **21**, 1003–1009.

Morse S, Bennett R and Ismael I (2003) Weighing the economic costs and benefits of Bt cotton in South Africa. ISB News Report May 2003, 1–4.

Morse, S, Bennett, R and Ismael, Y 2004 Why Bt cotton pays for small-scale producers in South Africa. *Nature Biotechnology* **22**, 379–380.

Munkvold GP, Hellmich RL and Rice LG (1999) Comparison of fumonisin concentrations in kernels of transgenic Bt maize hybrids and non-transgenic hybrids. *Plant Disease* **83**, 130–138.

Naranjo SE (2005) Long-term assessment of the effects of transgenic Bt cotton on the abundance of nontarget arthropod natural enemies. *Environmental Entomology* **34**, 1193–1210.

Nielsen KM and Townsend JP (2004) Monitoring and modeling horizontal gene transfer. *Nature Biotechnology* **22**, 1110–1114.

Pingali P (Ed.) 2001 *CIMMYT 1999/2000 World Maize Facts and Trends. Meeting World Maize Needs, Technological Opportunities and Priorities for The Public Sector.* (CIMMYT: Mexico.)

Pray CE, Huang J, Hu R and Rozelle S (2002) Five years of Bt cotton in China – the benefits continue. *The Plant Journal* **31**, 423–430.

Qaim M and Zilberman D (2003) Bt crops can have substantial yield effects. ISB News Report April 2003, 5–6 (quoting Qaim and Zilberman 2003 Yield effects of genetically modified crops in developing countries. *Science* **299**, 900–902.

Qaim M and de Janvry A (2003) Genetically modified crops, corporate pricing strategies, and farmers' adoption, the case of Bt cotton in

Argentina. *American Journal of Agricultural Economics* **85**, 814–828.

Romeis J, Meissle M and Bigler F (2006) Transgenic crops expressing *Bacillus thuringiensis* toxins and biological control. *Nature Biotechnology* **24**, 63–71.

Salyer T (2006) Most growers aware of complying with Bt corn standards. *ISB News Report* June 2006.

Schnepf E, Crickmore N, van Rie J, Lereclus D, Baum J, Feitelson J, Zeigler DR and Dean DH (1998) *Bacillus thuringiensis* and its pesticidal crystal proteins. *Microbiology and Molecular Biology Reviews* **62**, 775–806.

Shelton AM, Zhao J–Z and Rouch RT (2002) Economic, ecological, food safety and social consequences of the deployment of Bt transgenic plants. *Annual Review of Entomology* **47**, 845–881.

Stotzky G (2002) Release, persistence, and biological activity in soil of insecticidal protein from *Bacillus thuringiensis*. In *Geneically Engineered Organisms*. (Eds DK Letourneau and BE Burrows.) pp. 187–222. (CRC Press: London.)

Tabashnik B, Carrière Y, Dennehy TJ, Morin S, Sisterson MS, Roush RT, Shelton AM and Zhao J–Z (2003) Insect resistance to transgenic Bt crops, lessons from the laboratory and field. *Journal of Economic Entomology* **96**, 1031–1038.

Tabashnik B, Dennehy TJ and Carrière Y (2005) Delayed resistance to transgenic cotton in pink bollworm. *Proceedings of the National Academy of Sciences of the USA* **102**, 15389–15393.

Thirtle C, Beyers L, Ismael Y and Piesse J (2003) Can GM-technologies help the poor? The impact of Bt cotton in Makhathini Flats, KwaZulu-Natal. *World Development* **31**, 717–732.

Traxler G, Godoy-Avila S, Falck-Zepeda J and Espinoza-Areilano, J (2003) Transgenic cotton in Mexico, economic and environmental impacts. In *The Economic and Environmental Impacts of Agbiotech, A Global Perspective*. (Ed. N Kalaitzandonakes.) pp. 183–202. (Kluwer-Plenum Academic Publishers: New York.)

Van der Walt W (2006) GM maize doubles market share in 2006. *Nufarmer and African Entrepreneur* **10**, 21–25.

Vercesi ML, Krogh PH and Holmstrup M (2006) Can *Bacillus thuringiensis* (Bt) corn residues and Bt corn plants affect life-history traits in the earthworm *Aporrectodea caliginosa*? *Applied Soil Ecology* **32**, 180–87.

Wang J, Wang S, Su J, Huang T, Hu X, Yu J, Wei Z, Liang Y, Liu Y, Luo H and Sun G (2003) Food contamination of fumonsisn B_1 in high-risk area of esophageal and liver cancer. *Toxicological Science* **72**, 188–196.

Wu KM and Guo YY (2005) The evolution of cotton pest management practices in China. *Annual Review of Entomology* **50**, 31–52.

Zhao J–H, Cao J, Li Y, Collins HL, Roush RT, Earle ED and Shelton AM (2003) Transgenic plants expressing two *Bacillus thuringiensis* toxins delay insect resistance evolution *Nature Biotechnology* **21**, 1493–1497.

3

Herbicide-tolerant crops

Weeds are an ever-present part of any farmer's life. In the developed world most farmers control weeds by spraying with chemical herbicides or weedkillers. About 5% of all fossil fuel use is by agriculture, and most of this goes on weed control. In spite of this it is estimated that 12% of global crop yield is lost to weeds (Paoletti and Pimentel 2000). In the developing world farmers who cannot afford herbicides have to resort to manual weeding, a backbreaking job, usually undertaken by women.

Most herbicides target selected weeds and farmers choose one that is tolerated by the crop. For example, sugar beet varieties grown in the UK are tolerant to about 18 commercially available herbicides. By using a combination of up to eight different herbicides, applied at times when different weeds are a problem, farmers can usually protect their crops. Without such protection sugar beet yields would fall by about three-quarters and the crop would not be worth harvesting (Halford 2003).

There are some alternative weed control strategies such as in the case of rice cultivation. Historically rice has been grown under water as a method of weed control. Although this works well, it uses enormous quantities of fresh water. For example, in Australia, rice production accounts for 0.02% of the GDP but requires 7% of the fresh water used by the country. Little wonder then that biotechnology companies have developed herbicide-tolerant rice that can be grown on dry land (Bruce Chassy, pers. comm. 2006).

Plant breeders have for some time been trying to select herbicide-tolerant varieties of certain crops. Thus sulfonyl urea-tolerant soybean, imazamox-tolerant sunflower and imidazolinone-tolerant maize are all available. Therefore, the concept of herbicide-tolerant crops predates the advent of the use of genetic modification for tolerance.

The first GM herbicide-tolerant crop to be introduced in 1996 was Roundup Ready soybean. Roundup is Monsanto's trade name for **glyphosate**, a broad-range herbicide that is practically non-toxic to organisms other than plants. Unlike many other herbicides, it has very little toxicity to insects, birds, fish or mammals, including humans (Extoxnet 1996; Pesticide News 2003). It does not persist long in the soil as it is broken down by microorganisms within days or several months, depending on soil type. Glyphosate targets an enzyme involved in **amino acid** synthesis and therefore plants sensitive to Roundup are unable to make proteins, resulting in plant death. GM Roundup Ready crops are resistant to the activity of glyphosate.

Herbicide-tolerant crops are the most widely planted of all GM crops. In 2004, herbicide-tolerant soybean, maize, canola and cotton was planted on 72% of the 81 million hectares of these crops (James 2004).

Herbicide-tolerant GM soybeans accounted for 80% of the plantings in the USA in 2003, while worldwide the acreage increased by 13% compared with 2002 (James 2003). This increased by 17% in 2004 (James 2004). Argentina reported that almost all of its 14.5 to 15 million hectares of soybeans were planted with herbicide-tolerant varieties. Significant plantings were also reported for Brazil, Canada, Uruguay, Romania, South Africa and Mexico.

Transgenic canola is a crop grown over much smaller areas but it also showed an increase of 20% in global plantings in 2003. Interestingly, in Canada, 68% of canola was GM in 2003, with an additional 22% planted to herbicide-tolerant varieties derived by mutation (James 2003). In 2004 almost all the increase in plantings of GM canola occurred in Canada (James 2004).

Apart from ease of weed control on a given crop, another advantage of using glyphosate-tolerant varieties is that crop rotation is made much easier. Some of the herbicides used on conventional soybeans remain active through to the next season and beyond. Maize, typically grown in rotation with soybeans, requires an eight to 10-month gap after the application of some herbicides that are used on soybean, before it can be planted. There are no such problems after glyphosate treatment as this herbicide is degraded so rapidly in the soil (Halford 2003).

Glyphosate-tolerant varieties are not the only GM herbicide-tolerant plant varieties available. The herbicide **glufosinate** is marketed under the trade name of Liberty®. This active ingredient disrupts amino acid synthesis leading to the accumulation of toxic levels of ammonia. This causes **photosynthesis** to stop and the plant dies within a few days. GM Liberty Link-tolerant crops are resistant to the activity of glufosinate.

Glyphosate and glufosinate are far more environmentally friendly than many herbicides used on conventional crops. Thus not all herbicides are equal and if the use of glyphosate and glufosinate is increasing, this should be balanced by the decrease in the use of other herbicides.

Economic impacts of herbicide-tolerant crops

Brookes and Barfoot (2005) conducted an assessment of the economic impacts of herbicide-tolerant crops from the first year of planting, 1996, until 2004. These crops were not developed in order to increase yield, and therefore all farm income effects are associated with changes in the cost of production. These include the use of reduced- or no-tillage systems (see Environmental impacts of herbicide-tolerant crops), which allows, among other benefits, additional time for planting, growing and harvesting second crops (Table 3.1).

The largest gains in farm income arose in the soybean sector, largely from cost savings, where additional income was equivalent to adding 9.5% to the value of the crop in GM-growing countries, or adding the equivalent of 6.7% to the $62 billion value of the global soybean crop.

Table 3.1 Economic impacts of herbicide-tolerant crops from 1996 until 2004

Crop	Country	Cost savings ($/ha)
Soybeans	USA	25.2 in 1996 and 1997 33.9 1998–2000 73.4 in 2003 78.5 in 2004
	Argentina	24–30 all years depending on exchange rate
	Brazil, Paraguay and Uruguay	88 all years using 2005 exchange rate
	Canada	C$47–89 1997–2004
	Romania	140–239 1999–2003
Maize	USA	39.9 all years
	Canada	C$48.75 all years
Cotton	USA	34.12 1996–2000 66.59 2001 onwards
	Australia	A$60 all years from 2000
Canola	USA	60.75 1999–2001 67 2002 onwards for glyphosate tolerance 44.89 all years for glufosinate tolerance
	Canada	C$39 all years

From Brookes and Barfoot (2005).

Environmental impacts of herbicide-tolerant crops

In 2003 the UK Royal Society published a report covering a four-year study on the impact of herbicide GM crops on farm biodiversity (The Royal Society 2003). A brochure has also been produced by the independent Scientific Steering Committee to aid public understanding of the Farm Scale Evaluation (FSE) results (Defra 2006).

The FSE were held because of concerns that GM herbicide-tolerant crops might exacerbate the negative impacts upon British farmland wildlife already apparent over the last four decades as a result of intensification of farming practices. There was concern that control of weeds in herbicide-tolerant GM crops might be so efficient that their use would help to remove weeds from fields where they had previously grown causing a decline in the wildlife that depended on them. However, there were those who argued that these crops might reduce herbicide use and thus allow the growth of weeds and their associated wildlife.

The study began in 1999 and was the largest set of field trials of GM crops in the world. It involved 266 trial fields around England, Wales and Scotland. The crops studied were winter-sown and spring-sown canola (oilseed rape), sugar beets and maize. There was an approximately 36% reduction in the application of sprays on GM sugar beet and similar reductions in maize, but not with canola (Champion et al. 2003).

The environmental impact results showed that growing conventional beet and spring rape was better for many groups of wildlife than growing the GM varieties. There were more insects, such as butterflies and bees, in and around the conventional crops because there were more weeds to provide food and shelter. There were also more weed seeds in the conventional crops. Weed seeds are an important part of the diets of some animals, particularly farmland birds. Growing GM winter rape resulted in the same number of weeds as in the conventional rapeseed crop. However, in the GM crops there were fewer broad-leaved, flowering weeds as the herbicide used is specific for such weeds. There were fewer bees and butterflies but no differences in the numbers of other insects, slugs and spiders.

In contrast growing GM maize was better for many groups of wildlife. There were more weeds in and around these varieties, more butterflies and bees, and more weed seeds. This result was not surprising as the herbicide applied to the conventional maize plants was **atrazine**, a persistent, broad-spectrum herbicide, and a much more efficient weedkiller than glufosinate applied to the GM varieties. Indeed, conventional maize had the fewest weeds and produced the fewest weed seeds of all the crops tested. An interesting result was that all three GM crops significantly increased the numbers of insects feeding on decaying plant material (Champion et al. 2003).

The researchers stress that the differences they found did not arise because the crops had been genetically modified. They arise because these GM crops give farmers new options for weed control. They can use different herbicides and apply them differently.

Dr Les Firbank, coordinator of the project, stated:

> *'although the trials portrayed as a test of the environmental credentials of GM crops, it is really the weed killers to which they are resistant, that are being evaluated'* and *'if the aim is to save farmland wildlife, banning any of the GM crops tested, is unlikely to make any difference.'*

The study revealed significant differences in the management of the four herbicide-tolerant and conventional crops, and emphasised the importance of weeds growing among crop plants in sustaining natural communities in agricultural fields. But the question that was not answered is the following: Is the UK farming system there to promote weeds and the build-up of weed seeds or is it there to produce as much food from as little arable land as possible? Neither were the following addressed: Do herbicide-tolerant GM crops reduce the use of herbicides that persist in the environment and does the use of these crops promote no-tillage practices that benefit soil health?

In a letter to *Nature Biotechnology* entitled 'UK field-scale evaluations answer wrong questions' a group of food and agricultural scientists address some of these issues (Chassy et al. 2003). They pointed out that weed populations are a result of the farmer's weed management strategy, not the GM status of the crop.

For example, an organic farmer who thoroughly hoes a field would be equally effective at destroying potential feed and habitats for birds. A farmer using conventional herbicides together with mechanical tillage might do likewise. Indeed, the studies demonstrated that weeds and some insects were more common in oilseed rape crops, GM or conventional, than in beet or maize crops. It could therefore be argued that a more effective method of increasing the numbers of arable weeds and insects in crops would be to legislate crop choice.

The question of no-till practices is an interesting one. Conservation tillage practices have been in place for many years. Prior to their use farmers would till, or plough, the soil to kill weeds present and encourage the growth of new weeds. Once these emerged they would spray, usually with a type of herbicide which persisted for some time in the soil. Farmers would delay planting their seed until they estimated that enough of the herbicide had dispersed in order for it not to harm the germinating plants. During that period, precious top-soil might be lost due to wind and rain. Ploughing causes runoff, that pollutes and erosion that silts, rivers. It kills the earthworms and microbes that naturally till and nourish the soil. It burns diesel fuel that pollutes the air. Thus no-till or conservation tillage has great benefits to the environment and biodiversity. With the advent of herbicide-tolerant GM crops, farmers can increase their use of conservation tillage as once the weeds emerge they can spray without harming the crop itself. This no-till agriculture leads to a decrease in energy inputs, lower soil erosion and soils that are much healthier with respect to structure and microbes, invertebrates and organic matter content.

Chassy et al. (2003) suggest that to truly test the impact of the GM nature of a crop on biodiversity, instead of the effect of a cropping system, the trials should have focused on comparing a single GM crop with its carefully matched conventional counterpart. Thus, they should grow sulfonylurea-tolerant GM canola and conventionally bred sulfonylurea-tolerant plants, with and without sulfonylurea treatment. 'In all likelihood, the studies would have found little difference in biodiversity between the planting of GM and conventional sulfonylurea-tolerant cultivars.' Lack of spraying on either cultivar would have resulted in more weeds, weed seeds and insects regardless of its GM nature. 'Put in another way, these studies were not even about GM crops!'

Herbicide-tolerant technology, whether GM or not, may be one of those rare technologies that improves both the yield and product quality of a crop while reducing adverse environmental effects. By increasing productivity on existing farmland, surplus arable land could be conserved for natural reserves thereby providing even greater biodiversity.

In another analysis of the UK study, Dewar et al. (2004) commented that there are two approaches to the way in which society might want food to be produced:

1. A return to low intensity farming that would require more land to be brought into production to compensate for the reduced yields that would inevitably ensue; or

2. Further intensification on the most productive areas of land that would allow less productive areas to be returned to wildlife.

The first approach, which would be more popular with environmentalists than farmers, could use GM crops but apply the herbicide only once early in the season. This allows later emerging, but non-competitive weeds to mature and produce seeds. Of course the crops will look untidy, but a pristine crop is not necessarily an environmentally friendly one. However, the power of GM herbicide tolerance will give farmers the confidence that yields will not decrease even if some weeds remain. That could encourage them to adopt a low intensity approach.

The second approach, more acceptable to farmers but less to environmentalists, is based on the assumption that crops do need to be grown to produce food. Is it better to provide a larger area of poor diversity or many small areas of rich diversity? These questions remain to be tested.

Another study was published in 2004 on a life-cycle assessment (LCA) of growing GM herbicide-tolerant sugar beet in the UK and Germany (Bennett et al. 2004). LCA is an accepted method for assessing environmental and human health impacts associated with a product or process. It has been used for some time to analyse industrial products and processes but has only recently been applied to agriculture. LCA lists environmental burdens and analyses their impact in terms of their relative importance to the environment and human health. The results suggested that growing the GM variety would be less harmful than the non-GM variety to the environment and human health. This was largely as a result of lower emissions from herbicide manufacture, transport and field operation. Emissions that contribute to global warming, ozone depletion, water toxicity as well as acidification and nutrification of soil and water, were much lower for the GM crop. In addition, emissions that contribute to summer smog and toxic particles in the air and thus impact negatively on human health, were substantially lower for the GM crop.

In an analysis of GM sugar beet, scientists at the UK's Broom's Barn Research Station have shown that herbicide-tolerant varieties can be used to the benefit of the environment (May et al. 2005). To obtain wildlife benefits in spring, the authors improved the timing of the herbicide application to maximise both crop yields and the benefits derived from leaving weeds between crop rows. Autumn environmental benefits are more important as autumn weeds provide seeds for bird food and for recharging weed seed banks, as well as increasing the number of seeds left in the soil to grow into weeds. The paper demonstrates a system that gives maximum crop yield and increased weed seed availability of up to 16-fold. The new system is extremely simple. It involves applying the first spray fairly early and omitting the second spray. This gives additional cost and pesticide savings on top of the already large savings compared to conventional practice. According to May et al. (2005), conventional weed control does not have the flexibility to benefit both agriculture and the environment.

Herbicide-tolerant oilseed rape in Scotland – a farm-scale evaluation trial

Shirley Harrison, a farmer in the north-east of Scotland, presented a paper on her experiences with herbicide-tolerant canola at the CropLife International Conference in Brussels in 2003. The results she has seen have convinced her of the benefits of the technology. Despite having her fields vandalised three times, the GM half of the crop received just one herbicide treatment. In her words:

> *'The weed varieties were knocked back and destroyed very satisfactorily, enabling the crop to grow and establish a canopy. With the non-GM crop we were battling to control the wild oats with the target spray and while the general herbicide had checked the weeds, there was a lot of general weed rubbish under the canopy. The GM crop was clean ... I saved 84 pounds/ha by growing GM oilseed rape.'*

Shirley also noticed the prevalence of wildlife in the GM half of the crop, 'While small birds were busy on the GM half, they did not linger long in the conventional crop.'

The effect of herbicide-tolerant crops in developing countries

The above discussion on the effects of herbicide-tolerant crops is, of course, hardly relevant to most of the developing world where herbicides are simply beyond the means of most farmers. Indeed, many African women spend as much as 25% of their working life weeding. But herbicide-tolerant maize may hold the answer for one of their most pernicious weeds, *Striga hermonthica* or witchweed. This insidious parasitic weed, which produces pretty flowers, is extremely difficult to remove as it entwines its roots within the roots of the crop it attacks, such as maize, sorghum and millet throughout sub-Saharan Africa. Moreover, it especially targets the roots of poorly growing crops that might have been subjected to drought or pests.

The weed infests an estimated 20 to 40 million hectares of smallholder farmland and causes yield losses ranging from 20% to 80% and up to 100% in severe infestations. Before emerging from the soil it produces poisonous substances that are harmful to crops, and it drains food, minerals and water from its host. Striga seeds can remain dormant and viable in the soil for up to 20 years. With every planting season some of the dormant seeds, stimulated by crop exudates, grow, stunting the crop while producing more seeds that add to the dormant ones, worsening the problem.

What can be done? Scientists have bred non-GM lines of maize that are resistant to the herbicide **imazapyr**. They have also developed a method for coating maize seeds with a formulation of the herbicide. As the maize seeds germinate they release imazapyr over time and, when the witchweed attacks, it is killed. Field trials in Kenya have shown this to be most effective and the amount of herbicide used is minimal (Figure 3.1). The herbicide-coated hybrid maize seeds will be sold at a cost of US$4/ha more than normal hybrid seeds. As the benefits are esti-

Figure 3.1 A non-imazapyr-resistant stand of maize (left) compared with an imazapyr-resistant stand (right) (courtesy of the African Agricultural Technology Foundation).

mated to be US$52/ha the cost:benefit ratio should be 4:52. For farmers who don't normally buy hybrid seeds the cost:benefit ratio should be 26:52. It will be interesting to see how farmers perceive these benefits and whether they are prepared to buy the imazapyr-coated seeds.

The use of Roundup Ready soybeans in Argentina has raised several issues. There appears to have been a dramatic increase in the use of Roundup since the release of the GM seeds. In part this is a result of the expiry of Monsanto's patent on glyphosate. In 2001, 22 companies were able to provide generic versions of the product at competitive prices (Qaim and Traxler 2005). However, glyphosate has no residual activity and is rapidly decomposed by soil microorganisms. The increased use of glyphosate also significantly reduced applications of more hazardous herbicides in higher toxicity classes.

Because soybean production in Argentina is fully mechanised, the use of herbicide does not displace hand-weeding labourers. Whether the planting of herbicide-resistant crops in other developing countries will affect labour will depend on the type of agriculture practiced as well as on the type of crop. Weeding sometimes provides the labourers, mainly women, with their only source of income. However, the use of GM crops that reduce labour could significantly address specific social and economic crises facing rural communities as a result of the AIDS pandemic. Shortage of farm labourers often means that children are increasingly involved in agriculture, impacting negatively on their education and quality of life.

There is another issue of concern about the widespread use of herbicide-tolerant crops and that is the possibility of gene flow to potentially weedy relatives. Some have coined the term 'superweeds' to raise public awareness. This issue will be discussed in detail in Chapter 6.

Summary

Weeds are an ever-present part of any farmer's life. GM herbicide-tolerant varieties of several crops are being grown in increasing numbers worldwide. These are resistant to herbicides that are much more environmentally friendly than many weedkillers used on conventional crops. Farm-scale evaluations of the environmental impacts of GM herbicide-tolerant crops were carried out in the UK. Some were found to decrease the numbers of weeds and therefore the numbers of insects. Some were found to have little effect and some to have beneficial effects. However, this had nothing to do with the crops being genetically modified per se. Rather it reflected the altered use of herbicides. A more recent finding has been that effective use of herbicides on GM crops can increase the amount of wildlife in the fields. Another benefit of herbicide-tolerant GM crops is the decrease in tillage with resultant environmental benefits. Herbicide-tolerant crops in developing countries could have a major impact on labour practices.

References

Bennett R, Phipps R, Strange A and Grey P (2004) Environmental and human health impacts of growing genetically modified herbicide-tolerant sugar beet, a life-cycle assessment. *Plant Biotechnology Journal* **2**, 273–278.

Brookes G and Barfoot P (2005) GM crops, the global economic and environmental impact – the first nine years 1996–2004. *AgBioForum* **8**, 187–196.

Champion GT, May MJ, Bennett S, Brooks DR, Clark SJ, Daniels RE, Firbank LG, Haughton AJ, Hawes C, Heard MS et al. (2003) Crop management and agronomic context of the Farm Scale Evaluations of genetically modified herbicide-tolerant crops. *Philosophical Transactions of the Royal Society Biological Sciences* **358**, 1801–1818.

Chassy B, Carter C, McGloughlin M, McHughen A, Parrott W, Preston C, Roush R, Shelton A and Strauss SH (2003) UK field-scale evaluations answer wrong questions. *Nature Biotechnology* **21**, 1429–1430.

Defra (2006) *The Farm Scale Evaluations.* http://www.defra.gov.uk/ environment/gm/fse [Verified 19 April 2006].

Dewar AM, May MJ and Pidgeon JD (2004) Environmental impact of GM herbicide-tolerant crops, the UK farm scale evaluations and proposal for mitigation. ISB News Report February 2004. (Information Systems for Biotechnology: Blacksburg, VA.)

Extoxnet (1996) Glyphosate. http://extoxnet.orst.edu/pips/glyphosa. htm [Verified 2 June 2006].

James C (2003) Global status of commercialized trangenic crops. 2003 ISAAA Brief no. 30. (International Service for the Acquisition of Agri-biotech Applications: Ithaca, NY.)

James C (2004) Global status of commercialized biotech/GM crops. 2004 ISAAA Brief no. 32. (International Service for the Acquisition of Agri-biotech Applications: Ithaca, NY.)

Halford N (2003) *Genetically Modified Crops*. pp. 41–45. (Imperial College Press: London.)

May MJ, Champion GT, Dewar AM, Qi A and Pidgeon JD (2005) Management of genetically modified herbicide-tolerant sugar beet for spring and autumn environmental benefit. *Proceedings of the Royal Society B* **272**, 111–119.

Paoletti M and Pimentel D (2000) Environmental risks of pestidices versus genetic engineering for agricultural pest control. *Journal of Agricultural and Environmental Ethics* **12**, 279–303.

Qaim M and Traxler G (2005) Roundup Ready soybeans in Argentina, farm level and aggregate welfare effect. *Agricultural Economics* **32**, 73–86.

Pesticide News (2003) http://www.pan-uk.org/pestnews/actives/glyphosa.htm. [Verified 19 April 2006].

The Royal Society (2003) Philosophical Transactions of The Royal Society Biological Sciences 29 November 2003 **358 (1439)**, 1775–1913 http://www.journals.royalsoc.ac.uk/(lp3uekyorzencc45w4mtz52j)/app/home/search-articles-results.asp?referrer=main [Verified 2 June 2006].

4

Virus-resistant and drought-tolerant crops

Crops that are more resistant to virus infection by virtue of being genetically modified are not widely planted. However, they could have a significant impact in areas of the world where plant viruses are locally widespread. Because many plant viruses are spread by vectors, usually insects, the impact of these viruses occurs in regions where both are prevalent. One example is Maize streak virus (MSV), spread by a leafhopper, *Cicadulina mbila* and related species, and found almost exclusively on the African continent and neighbouring islands such as Madagascar. Another is Cassava mosaic virus (CMV), spread by whiteflies (*Bemisia tabaci*) and again it is common in countries on the African continent, such as Uganda, where cassava (also known as manioc) is grown as a staple crop. Indeed, in the 1990s, Uganda nearly lost its entire cassava crop to CMV. The virus is spreading northwards and agriculturalists in Nigeria, one of the world's largest producers of cassava, are becoming extremely concerned about this.

Virus-resistant papaya

Papaya (*Carica papaya*) has been grown in Hawaii for many years. The Papaya ringspot virus (PRSV) is transmitted by aphids and is a major problem, both on the Hawaiian islands and in other tropical regions where the crop is grown. Indeed, in 1994 the Hawaiian industry found itself in a crisis. As a result, a research team, led by Dennis Gonsalves, set about developing commercial varieties of papaya resistant to PRSV (Gonsalves 2004).

The strategy they used was based on the genetic material of the virus being **RNA**, not DNA, encased in a coat of identical copies of a protein, the coat protein. When the virus infects a plant, it 'uncoats' itself, replicates its RNA, spreads through the plant, makes more coat proteins and eventually 're-coats' itself to be picked up by aphids and transferred to its next plant target. If the plant makes its own copies of the PRSV coat protein, the virus is never able to uncoat itself. Therefore, it cannot replicate its RNA and cannot spread through the plant. Thus the plant is resistant to the virus. Although this worked in some instances, broader resistance was obtained when the coat protein gene was only transcribed into RNA but not translated into protein (Gonsalves 1998). This worked extremely well in the field (Figure 4.1)

Figure 4.1 An aerial view of transgenic papaya plants surrounded by severely infected non-transgenic plants (Gonsalves 2004).

However, the regulatory process took some time. Eventually, in 1998, seeds were distributed free to growers, seven years after the resistant line was developed. Harvesting started in 1999, grower, packer and consumer acceptance were widespread and the papaya industry was 'spared from disaster' (Gonsalves 2004).

Virus-resistant potatoes

Potatoes (*Solanum tuberosum*) are plagued by several viruses including Potato X and Y (PVX and PVY), as well as Potato leafroll virus (PLRV). Commercialisation of the GM-crop was stopped by scientists in response to activities of the anti-GM crop activists. Plants with resistance to all three viruses had been developed using a strategy based on the genetic material of the virus being RNA, not DNA, encased in a coat of identical copies of a protein, the coat protein. The mechanism is the same as outlined in Virus resistant papaya on page 54.

Instead the constructs were donated to CINVESTAV (Centro de Investigaión y Estudios Avanzados, Mexico) to be used in local potato varieties by subsistence farmers (Kaniewski and Thomas 2004). It will be of interest to follow whether or not the anti-GM crop activists have acted to benefit small-scale farmers in Mexico.

Potatoes resistant to PVY were, however, commercialised in the USA and Canada, at great benefit to seed potato growers. The anti-biotechnology campaign against GM potatoes resulted, however, in McDonald's deciding to ban them from its food chain. Potato processors, under pressure from McDonald's and the European export markets, suspended their transgenic contracts. Despite this the potato virus project in Mexico has led to the development of virus-resistant transgenic lines of locally popular varieties of potatoes. In addition, trials are in progress in Russia, Bulgaria, Romania and Mauritius. In almost all cases, the advantages will be for small-scale subsistence farmers who cannot afford the use of pesticides to kill the insect vectors that transmit the viruses.

Virus-resistant maize

In her review on agriculture in the developing world, Delmer (2005) points out the importance of the development of virus-resistant crops specifically targeted to helping small-scale farmers in these regions.

Maize streak virus (MSV) is a major problem in many African countries. Breeders in Africa have succeeded in developing resistant maize varieties, but the resistance often 'breaks down' as a result of the development of more virulent virus strains. In addition, varieties bred for resistance in one African country are often not effective in others due to environmental and other conditions. It would, therefore, be advantageous to develop a transgenic maize line with a single gene conferring MSV resistance. This could then be crossed into regionally important commercial varieties.

The author's laboratory, together with that of Professor EP Rybicki, has been involved in developing such a line. Unlike the papaya and potato viruses mentioned above, MSV has DNA, not RNA, as its genetic material. For various reasons, the introduction of the coat protein gene will not protect a plant against infection from a DNA virus. The gene has been used, therefore, for another MSV protein, one involved in its replication, to develop transgenic virus-resistant maize. The results are very promising and field trials are planned for the 2006/07 growing season (Figure 4.2).

Virus-resistant cassava

Cassava mosaic disease (CMD) caused by Cassava mosaic virus (CMV) is the most important constraint to the production of cassava (or manioc, *Manihot esculenta*) in Africa (Legg and Fauquet 2004). A severe epidemic of CMD spread through Uganda in the 1990s causing losses valued at more than US$60 million annually between 1992 and 1997. Farmers abandoned the crop in many parts of the country resulting in food shortages. Later in the 1990s the epidemic spread to Sudan, Kenya, Tanzania and the Democratic Republic of Congo. Since then the disease has been managed to some extent by phytosanitation, comprising all the

Figure 4.2 Maize infected with Maize streak virus (MSV) (left) and virus-resistant transgenic maize infected with MSV (right).

techniques that can keep plants in a virus-free condition. This includes removal of CMD-diseased plants from within a field and the selection of symptom-free cassava stems for planting. There has also been some success in breeding CMV-resistant cultivars. However, the problem persists and scientists are developing GM lines of CMV-resistant cassava. As with MSV, CMV is a DNA virus and scientists are using a variety of strategies to develop crop resistance. Initial results are very promising (Chellappan et al. 2004).

Environmental impact of resistance to viruses

Viruses are able to swap their genetic material with related viruses. This process is called genetic recombination. This can potentially increase the virulence of a virus. The question is, therefore, can a GM plant, growing in the field and expressing a viral coat or replication protein, allow an infecting virus to recombine with these genes? If so, could the resultant virus be more virulent or have a broader host range? A six-year study was undertaken to answer these questions in Potato leafroll virus (Thomas et al. 1998). There were no significant changes in any viral characteristics during that time. It became clear that viruses commonly recombine with each other in the field and the presence of transgenic coat protein or replication genes make no significant detectable difference.

Drought-tolerant crops

Water is a major limiting factor in world agriculture with most crops being highly sensitive to even mild dehydration. About one-third of the 1.5 billion hectares of the world's arable land is affected by drought (James 2002). Many research groups are working on the development of crops such as maize and wheat that are tolerant to drought.

A research team at the University of Cape Town, including the author's laboratory, is using genes isolated from a South African indigenous 'resurrection plant' to develop drought-tolerant maize. Resurrection plants are unique in that they are able to tolerate almost complete desiccation. They can lose 95% of their water content and remain in a dormant stage, looking completely dead, for months on end. Upon the addition of water, the plants can literally 'resurrect' in a matter of days.

The resurrection plant being worked on (*Xerophyta viscosa*) is found growing in cracks in rocks in the Drakensberg mountains in the KwaZulu-Natal province of South Africa. The plants therefore grow in very little soil, and can dry out rapidly. In addition, day temperatures are often as high as 40°C while temperatures at night can drop to below freezing. These plants must have some genes coding for very interesting proteins to enable them to flourish in this environment.

This is just what was found. Among the genes that have been isolated are ones coding for proteins that protect the plant cell from water loss by replacing the water with sugars such as sorbitol. Others

Figure 4.3 (a) A hydrated resurrection plant, *Xerophyta viscosa*.

Figure 4.3 (b) A dehydrated plant.

bind to the cell membrane and probably act to detect dehydration stress and signal to the interior of the cell that other genes must be 'switched on' to produce proteins to protect the cell. Some of the latter proteins act as antioxidants to protect the DNA and other molecules in the cell from damage (Mundree et al. 2002).

Tobacco was used as a model plant to understand the drought-tolerance traits. The first transgenic tobacco plants produced expressed a membrane protein and the plants were tolerant to dehydration, high temperature and salt. This gene is now being tested in maize. However, nature is unlikely to use a single gene to produce drought tolerance. Therefore, more than one gene is being introduced into maize. This process is called stacking. The membrane protein gene is being stacked together with a gene that results in the accumulation of a sugar and one that produces an antioxidant.

Figure 4.4 Tobacco plants (*Nicotiana tabacum*) expressing a membrane protein from the resurrection plant, *Xerophyta viscosa*, and showing resistance to water stress. Non-transgenic plants on the left are almost completely dead. (Source, Dahlia Garwe.)

However, plants that produce these proteins throughout their life cycle are often phenotypically abnormal with their growth being stunted. Therefore, we must 'instruct' the plant to produce the proteins only when they are needed (e.g. during drought). These instructions are being introduced together with the genes. Other laboratories such as those of Kazuo Shinozaki in Japan (Kasuga et al. 2004) and Ray Wu in the USA (Su et al. 1998) have shown that such instructions can lead to plants that are able to tolerate some level of dehydration.

Several researchers have shown that the accumulation of the sugar trehalose confers resistance to a number of **abiotic stresses** in transgenic rice (Garg et al. 2002; Jang et al. 2003). The sugar helps stabilise biological molecules and protects against tissue damage during dehydration. Garg and others plan to seek patent protection for the modification resulting in trehalose accumulation and will ensure public availability of the modified crop to farmers in developing countries (Nuffield 2004).

Summary

Virus-resistant papayas have had a major impact on the papaya industry in Hawaii. Virus-resistant potatoes have not been widely grown as a result of the activities of anti-biotechnology lobbyists, while virus-resistant maize and cassava are still in the developmental stages in Africa. Similarly, drought-tolerant crops are in developmental stages.

References

Chellappan P, Masona MV, Vanitharani R, Taylor NJ and Fauquet CM (2004) Broad spectrum resistance to ssDNA viruses associated with transgene-induced gene silencing in cassava. *Plant Molecular Biology* **56**, 601–611.

Delmer DP (2005) Agriculture in the developing world, Connecting innovations in plant research to downstream applications. *Proceedings of the National Academy of Sciences of the USA* **102**, 15739–15746.

Garg AK, Kim J–K, Owens TG, Ranwala AP, Choi YD, Kochian LV and Wu, RJ (2002) Trehalose accumulation in rice plants confers high tolerance levels to different abiotic stresses. *Proceedings of the National Academy of Sciences of the USA* **99**, 15898–15903.

Gonsalves D (1998) Control of papaya ringspot virus in papaya, a case study. *Annual Review of Phytopathology* **36**, 415–437.

Gonsalves D (2004) Transgenic papaya in Hawaii and beyond. *AgBioForum* **7**, article 7 http://www.agbioforum.org/v7n12/v7n12a07-gonsalves.htm [Verified 20 April 2006].

James C (2002) Preview, Global status of commercialized transgenic crops, 2002. Brief No. 27. ISAAA p. 19. (International Service for the Acquisition of Agri-Biotech Applications: Ithaca, NY.)

Jang I–C, Oh S–J, Seo J–S, Choi W–B, Song SI, Kim CH, Kim YS, Seo H–S, Choi YD, Nahm BH and Kim J–K (2003) Expression of a bifunctional fusion of the *Escherichia coli* genes for trehalose-6-phosphate synthase and trehalose-6-phosphate phosphatase in transgenic rice plants increases trehalose accumulation and abiotic stress tolerance without stunting growth. *Plant Physiology* **131**, 516–524.

Kaniewski WK and Thomas PE (2004) The potato story. *AgBioForum* **7**, article 8 http://www.agbioforum.org/v7n12/v7n12a08-kaniewski.htm [Verified 19 April 2006].

Kasuga M, Miura S, Shinozaki K and Yamaguchi-Shinozaki K (2004) A combination of the *Arabidopsis* DREB1A gene and stress-inducible *rd29A* promoter improved drought- and low-temperature stress tolerance in tobacco by gene transfer. *Plant and Cell Physiology* **45**, 346–350.

Legg JP and Fauquet CM (2004) Cassava mosaic geminiviruses in Africa. *Plant Molecular Biology* **56**, 585–599.

Mundree SG, Baker B, Mowla S, Peters S, Marais S, Vander Willigen C, Govender K, Maredza A, Muyanga S, Farrant JM et al. (2002) Physiological and molecular insights into drought tolerance. *African Journal of Biotechnology* **1**, 28–38.

Nuffield (2004) *The Use of Genetically Modified Crops in Developing Countries.* (Nuffield Council on Bioethics, London UK.)

Su J, Shen Q, Ho, T–H D and Wu R (1998) Dehydration-stress-regulated transgene expression in stably transformed rice plants. *Plant Physiology* **117**, 913–922.

Thomas PE, Hassan S, Kaniewski WK, Lawson EC and Zalewski JC (1998) A search for evidence of virus/transgene interactions in potatoes transformed with the potato leafroll virus replicase and coat protein genes. *Molecular Breeding* **4**, 407–417.

5

Effects on biodiversity

A major concern about the introduction of GM crops into the environment is that such crops will affect and/or destroy biodiversity. Conner et al. (2003) commented that this has 'mainly contributed to more forests being destroyed in order to produce the paper on which the arguments have been presented'. As they point out, the discussion should better focus on whether GM crops pose threats to biodiversity that are different from those of conventional crops. Thus we need to compare the perceived benefits and potential threats of GM crops with the conventional ones they might replace.

It is important to view loss of biodiversity in a rational manner and ask what are the actual causes and what are the broader, less obvious, threats? The standard answers include climate change, deforestation and increase in agricultural land, but how many have considered the effects of wars? In an article in *New Scientist*, Fred Pearce (2005) describes how the mujahadeen fighters destroyed the national seed collection in Kabul in 1992. Afghanistan is the genetic heartland of several globally important crops. After the destruction of the Kabul collection, scientists at the International Centre for Agricultural Research in Dry Areas

(ICARDA) returned to the field, collected some key samples and hid them in the basements of houses before fleeing the country. When they returned they discovered that these had also been looted.

Similar biodiversity losses resulting from wars have occurred in Cambodia, the home of rice, the 'fertile crescent' in the Middle East, including Iraq, the Democratic Republic of Congo and Rwanda.

Klaus Ammann, head of the Botanical Garden in Bern, Switzerland, published an important review on the effects of GM crops on biodiversity (Ammann 2004). He followed this up with an article in 2005 (Ammann 2005). Much of this chapter comes from these articles. He stated that GM crops could have positive impacts on biodiversity including enabling growers to use fewer pesticides and decrease tillage. Ammann questions the widespread belief that GM crops lead inevitably to a reduction of genetic variability in crops. On the contrary, studies have shown that the genetic diversity can be increased. For instance, in the case of Roundup Ready (herbicide-resistant) soybean (see Chapter 3), the gene has been introduced into at least 400 varieties for cultivation in a wide range of environments. In addition, 'by slowing the rate at which natural habitats are destroyed, GM crops and other technologies that increase agricultural productivity can help to preserve the natural biodiversity'.

Biological diversity is a broad term, with different interpretations, that may refer to diversity within a given gene, a species, a community of species or an **ecosystem**. It is often contracted to the word biodiversity and used broadly with reference to the total biological diversity in an area or even the earth as a whole. Biodiversity comprises all living beings, from the simplest forms of viruses to the most sophisticated and highly evolved plants and animals (Ammann 2003).

Genetic, species and ecosystem diversity
Genetic diversity
Genes are the basic building blocks of life and are what make you different from me and both of us different from a maize plant. However, many genes are highly conserved across species. Humans share about 50% of

their genes with bananas and more than 98% with chimpanzees. The importance of genetic diversity is that combinations of genes within an organism, known as its genome, produce different physical effects that can affect the organism's resilience and survival under different environmental conditions. We do not know what untapped resources lie in the genomes of organisms, particularly in natural ecosystems. Therefore, it is prudent to retain and, if necessary, manage such ecosystems to preserve their genetic diversity.

Species diversity

Species are, for most practical purposes, the most appropriate targets for the measure of biodiversity. About 1.75 million species have been described but most, mainly insects, nematodes, fungi and bacteria, remain to be described scientifically.

Ecosystem diversity

Ecosystems can be classified as follows:

- Natural ecosystems that are free of human activities. It is debatable whether any true natural ecosystems exist on earth today.

- Semi-natural ecosystems in which human activity is limited. These are important ecosystems that are subject to some level of low intensity human disturbance. They typically abut managed ecosystems.

- Managed ecosystems are managed to varying degrees of intensity from the most intensive agriculture and urbanised areas, to less intensively managed systems. The latter includes some of the agricultural land in developing countries.

All ecosystems are important and we must work to preserve the broadest range of ecosystem diversity.

Loss of biodiversity and why worry?

Loss of biodiversity is occurring in many parts of the globe, often at a rapid rate. The threats can be ranked as follows:

1 habitat loss

2 introduction of exotic species

3 flooding/lack of water, climate changes.

Habitat loss as a result of the expansion of human activities has been identified as a main threat to 85% of all species. The main factors are urbanisation and the increase in cultivated land surfaces. In the year 2000, arable and permanent cropland covered about 1500 million hectares of land, with more than twice that amount classified as permanent pasture. This represents nearly 40% of the world's total available land surface (Ammann 2003).

So why should we worry about biodiversity? Biological diversity provides a source of significant economic, aesthetic, health and cultural benefits. The well being and prosperity of earth's ecological state as well as human society directly depend on the extent and status of biodiversity. In agriculture, biodiversity represents a critical source of genetic material allowing the development of new and improved crop varieties. There are also enormous, less tangible benefits including the values attached to the persistence of natural landscapes and wildlife, which become ever more important as these become scarcer.

In 1993 the Convention on Biological Diversity (CBD) came into force under the auspices of the United Nations Environment Programme (UNEP). It has three goals:

1 conservation of biodiversity

2 sustainable use of the components of biodiversity

3 fair and equitable sharing of benefits arising out of the use of genetic resources.

The most important change brought about by the CBD is the recognition that states have a sovereign right over the biodiversity within their territory. Previously it was considered that all organisms were the common heritage of everyone on earth. Now, under the terms of the CBD, living organisms or their products may be removed from a country only under mutually agreed conditions. Modern biotechnology has the potential to contribute to achieving these goals (see Cartagena Protocol on Biosafety).

The impacts of agriculture on biodiversity

Impacts on species biodiversity: agricultural biodiversity

Modern agricultural practices have been broadly linked to declines in biodiversity (Duelli et al. 1999). This can be found in many species including insects and birds. The extent of this decline depends on factors including cropping patterns, frequency of tillage, and the amount and nature of pesticides used, especially insecticides.

Cropping patterns

Intensive agriculture typically limits the diversity of crops grown. Many such systems are monocultures, at least in individual fields, and are relatively homogenous even at the regional level. Robinson and Sutherland (2002) analysed changes in agriculture and biodiversity in Britain since the 1940s. They found a consistent reduction in landscape diversity, reflected in a 65% decline in the numbers of farms. This was also associated with the removal of 50% of hedgerows, an important source of food and shelter for wild plants and animals.

Tillage

Intensive tillage leads to frequent disturbances of the agricultural landscape, increases problems of soil erosion and run-off from agricultural fields. This adversely affects the quality of agricultural habitats, with significant consequences for biodiversity.

Pesticide use

Conventional insecticides generally reduce diversity through direct toxic effects. Many of the widely used classes of conventional insecticides, including organophosphates and pyrethroids, have been shown to adversely affect a broad range of non-target species, including species of economic importance (Persley 2002; Ammann 2004). These include predator and parasitoid species, which keep crop pests in check. Local extinctions of these natural enemies are common where insecticides are frequently used, resulting in flare-ups of pests, some of which were not previously economically important. In a few cases, insecticides directly stimulate the population growth of non-target pest species (e.g. pyrethroids have such an effect on some mite and aphid species; Smith and Stratton 1986), which can then damage crops. Replacing broad-spectrum insecticides with more specific alternatives is necessary to avoid these impacts.

The impact of GM crops on species biodiversity in agriculture

The use of broad-spectrum insecticides can be significantly decreased by the adoption of insect-resistant Bt crops that express highly specific Bt proteins. This specificity means that very few non-target insects that do not feed directly on the crop plants will be adversely affected. Studies in which species diversity has been compared between Bt and non-Bt maize and cotton crops have indicated no significant differences (e.g. Lozzia 1999; Dively and Rose 2002; Fitt and Wilson 2003).

The only species that were observed to be significantly and consistently less abundant in fields of Bt crops are the target pests. In studies where the conventional crops were sprayed for that pest, many non-target species have been adversely affected.

Can Bt toxins persist in the soil and affect soil bacteria and other organisms? Bt toxins can certainly be released into the **rhizosphere**, which is the air and soil directly adjacent to the roots of plants. In addition binding tightly to clays in the soil stabilises the toxins. However, the toxins used in GM crops do not appear to have any consistent effects on earthworms, nematodes, fungi and other soil organisms (Saxena and Stotzky 2005).

Herbicide-tolerant crops may have a positive impact on species biodiversity due to the shift toward reduced tillage (see Chapter 3).

Impacts on species biodiversity: natural biodiversity

As discussed above, the factor most responsible for decreasing natural biodiversity is habitat destruction as land is cleared for agriculture. Since the early 1970s there has been a 1.68-fold increase in the amount of irrigated cropland and a 1.1-fold increase in cultivated land (Tilman 1999). This problem is most severe in developing countries with large levels of subsistence agriculture.

Aerial drift and transfer of insecticides in water can expose natural communities to potentially toxic levels of pesticides. In addition the impacts of tillage are even greater on natural habitats than on agricultural fields. Soil erosion as a result of tillage leads to high levels of fertilisers and pesticides being carried off agricultural fields into waterways. Nitrogen and phosphorus in fertilisers can have direct toxic effects on natural communities (Tilman 1999). Moreover, **eutrophication** leads to direct losses in biodiversity, pest outbreaks and changes in the structure of natural communities.

The impact of GM crops on natural species biodiversity

GM crops have the ability to benefit natural biodiversity in several ways. GM crops can increase yields thereby reducing the need to put additional land into agricultural production. In addition, insect-resistant crops reduce the use of broad-spectrum insecticides that could have direct and indirect effects on natural communities, especially those living close to agricultural fields. Herbicide-tolerant crops faciltitate decreased tillage, thereby reducing soil erosion, eutrophication and contamination of aquatic communities (see Chapter 3).

Impacts on genetic diversity: crop genetic diversity

Conventional agriculture relies on highly productive varieties generated through selective breeding programs. This has inevitably resulted in low crop genetic diversity, which has been the motivating force in the establishment of seed banks for important crops species in, for instance, institutes belonging to the Consultative Group on International Agricultural Research (CGIAR) system.

Concern has been expressed that the overall genetic diversity within crops will decrease as a result of genetic modification. Studies have shown that this is not the case. For example, Sneller (2003) (quoted in Ammann 2003) found that the introduction of herbicide tolerance into elite lines of soybeans in North America had little effect on soybean genetic diversity because of the widespread use of this trait in many different breeding programs. Similarly, Bowman et al. (2003) showed that genetic uniformity in cotton varieties in the USA had not changed significantly with the introduction of transgenic cotton cultivars. If anything, the genetic diversity had increased over the period of introduction of transgenic cultivars.

Impacts on genetic diversity: natural genetic diversity

Most biodiversity is concentrated in free-ranging habitats within the tropics between latitudes 23°26′ north (Tropic of Cancer) and 23°26′ south (Tropic of Capricorn). Thus, the greatest threat to biodiversity in these regions is the destruction and deterioration of habitats, as well as the introduction of invasive exotic species, particularly in tropical developing countries. In contrast, in temperate zones, particularly in the European Union, almost 50% of the landscape is agricultural. Therefore, agricultural land contains a significant portion of temperate-zone biodiversity. It is important to understand the differences in the impact of biodiversity in the world's tropical and temperate zones.

Biodiversity is impacted directly or indirectly by the needs of agricultural production. Increasing human population and limited arable land have demanded increased productivity leading to more intensive agricultural practices. In response to these demands, higher yielding crop varieties have been coupled with increased inputs such as fertilisers, pesticides and more intensive practices such as soil tillage. In a world where such practices are firmly entrenched as a vital part of food security, it is important to consider how GM crops can affect these, both positively and negatively. Overall, creating agricultural systems with minimal impact on biodiversity will require utilising all available technologies while simultaneously encouraging appropriate farming practices.

Intensive agriculture has negative direct impacts on the biodiversity of different species and of genetic diversity within a species. This is due to the use of pesticides and tillage (different species) and the practice of growing uniform crop varieties (diversity within a species). These impacts can be addressed by reducing broad-spectrum insecticides and tillage, and by encouraging diversification of agricultural systems. In contrast, many indirect agricultural impacts on natural biodiversity stem primarily from the conversion of natural habitats into agricultural production.

The Cartagena Protocol on Biosafety

- The Cartagena Protocol on Biosafety is part of the Convention on Biological Diversity (CBD). The CBD's three goals have been listed in 'Loss of biodiversity and why worry?'.

Modern biotechnology has the potential to contribute to achieving these three goals, as long as it is developed and used with adequate safety measures for the environment and human health. Thus the biosafety protocol, named after the Colombian city where the final round of treaty talks was launched in January 2000, is designed to enable people everywhere to enjoy the benefits of biotechnology while avoiding unnecessary risks. It sets out a comprehensive regulatory system for ensuring the safe transfer, handling and use of genetically modified organisms (GMOs) (or living modified organisms, LMOs) especially when they are being transported from one country to another.

At a practical level the protocol has two separate sets of procedures. One is for LMOs that are to be intentionally introduced into the environment, such as the planting of seeds for insect-resistant or herbicide-tolerant crops. The other is for LMOs that are to be used directly as human food, animal feed or for food or feed processing. This would include soybeans or maize kernels not to be used for planting.

The most rigorous procedures are reserved for the first case. The exporter gives the government of the importing country detailed written information in advance. A competent national authority must be in place in the importing country to authorise or reject the shipment. If no such authority exists in a country, no export to it is possible. This ensures that recipient countries have the opportunity to assess any risks that may be associated with a GMO before agreeing to its importation.

The latter case covers the largest category of LMOs, namely the shipment of GM maize, soybeans and other agricultural commodities intended for direct use as food, feed or processing, and not as seeds for growing new crops. There is, therefore, a simpler system for these imports. If a government decides to approve these commodities for domestic use, it has to communicate this decision through the Biosafety Clearing-House. This is one of the cornerstones of the Protocol. It is an internet-based system (CBD 2005). Common formats are used to ensure that the information collected from different countries is comparable.

The protocol allows governments to decide whether or not to accept imports of GMOs on the basis of risk assessments. These identify and evaluate potential adverse effects that a GMO may have on the environment, human or animal health. Although the country considering the import is responsible for ensuring that a risk assessment is carried out, it has the right to require the exporter to do the work and bear the cost. This is particularly important for developing countries.

In order to help developing countries establish competent national authorities the United Nations Environment Programme (UNEP), together with the Global Environment Facility (GEF), set out in 1997 to assist countries prepare National Biosafety Frameworks (NBF). In Africa, four countries, Kenya, Uganda, Cameroon and Namibia, are putting their draft Frameworks into action, 19 have published drafts and 41 are preparing theirs (UNEP-GEP Biosafety Projects 2006). As of January 2006, 57 countries worldwide have completed most of their Development of National Biosafety Projects and have draft NBFs on the website and many others are well advanced. These projects have generated a wealth of in-country experience in building capacity for biosafety.

In conclusion, where GM crops are adapted to new environments, especially marginal agricultural conditions such as salinity or drought, indigenous plants may be at risk. But this risk is no greater than it would be for similar crops bred by traditional methods. Let us not forget that by adapting GM crops to such marginal areas, biodiversity could be conserved. A major indirect threat to biodiversity is the loss of habitats due to the conversion of natural ecosystems to agricultural use in response to food demands. GM crops that could give higher yields on poor soils may alleviate the threat of habitat loss and thus contribute to biodiversity. Waste in rich countries and population growth in poor ones put great pressure on biodiversity. GM crops are no more or less likely to add to this pressure than any other agricultural development. 'It will largely be the social-economic and political context of the application of genetic modification that will determine whether the perceived threats or potential benefits of GM crops on biodiversity become a reality' (Conner et al. 2003).

Summary

Will GM crops affect biodiversity adversely? On the contrary, there is evidence that GM crops can actually contribute positively to genetic diversity. In contrast, many standard agricultural practices, such as cropping patterns, tillage and pesticide use may well have a greater negative impact on biodiversity. A major indirect threat to biodiversity is the loss of habitats due to the conversion of natural ecosystems to agriculture. GM crops may enable farmers to decrease their use of such marginal land for agricultural purposes.

References

Ammann K (2004) The impact of agricultural biotechnology on biodiversity, a review. http://www.botanischergarten.ch/Biotech-Biodiv/Report-Biodiv-Biotech12.pdf [Verified 5 June 2006].

Ammann K (2005) Effects of biotechnology on diodiversity, herbicide-tolerant and insect-resistant GM crops. *Trends in Biotechnology* **23**, 388–394.

Bowman DT, May OL and Creech JB (2003) Genetic uniformity of the US upland cotton crop since the introduction of transgenic cottons. *Crop Science* **43**, 515–518.

CBD 2005 Biosafety Clearing-House. http://bch.biodiv.org [Verified 20 April 2006].

Conner AJ, Glare TR and Nap J-P (2003) The release of genetically modified crops into the environment. Part II. Overview of ecological risk assessment. *The Plant Journal* **33**, 19–46.

Dively GP and Rose R (2002) Effects of Bt transgenic and conventional insecticide control on the non-target invertebrate community in sweet corn. In *Proceedings of the First International Symposium of Biological Control of Arthropods.* (US Forest Service: Morgantown, USA.)

Duelli P, Obrist MK and Schmatz DR (1999) Biodiversity evaluation in agricultural landscapes, above-ground insects. *Agriculture, Ecosystems and Environment* **74**, 33–64.

Fitt G and Wilson L (2003) Non-target effects of Bt-cotton, a case study from Australia. In *Biotechnology of* Bacillus thuringiensis *and its Environmental Impact.* (CSIRO Entomology: Melbourne.)

Head G, Surber JB, Watson JA, Martin JW and Duan JJ (2002) No detection of Cry1Ac protein in soil after multiple years of transgenic Bt Cotton (Bollgard®) use. *Environmental Entomology* **31**, 30–36.

Lozzia GC (1999) Biodiversity and structure of ground beetle assemblages (Coleoptera, Carabidae) in Bt corn and its effects on non target insect. *Bollettino di Zoologia agraria e di Bachicoltura Serie* 11 **31**, 37–58.

Pearce F (2005) Return to Eden. *New Scientist* **185**, 35–37.

Persley GJ (2003) Agricultural biotechnology, global challenges and emerging science. In *Agricultural Biotechnology, Country Case Studies – A Decade of Development.* (Eds Persley GJ and MacIntyre LR.) pp. 3–37. (CABI Publishing: Wallingford, UK.)

Robinson RA and Sutherland WJ (2002) Post-war changes in arable farming and biodiversity in Great Britain. *Journal of Applied Ecology* **39**, 157–176.

Saxena D and Stotzky G (2005) Release of larvicidal Cry protein in root exudates of transgenic Bt plants. *Information Systems for Biotechnology* **February 2005**, 1–3.

Smith TM and Stratton GW (1986) Effects of synthetic pyrethroid insecticides on nontarget organisms. *Residue Review* **97**, 93–120.

Sneller CH (2003) Impact of transgenic genotypes and subdivision on diversity within elite North American soybean germplasm. *Crop Science* **43**, 409–414.

Tilman D (1999) Global environmental impacts of agricultural expansion, sustainable and efficient practices. *Proceedings of the National Academy of Sciences USA* **96**, 5995–6000.

UNEP-GEP (2006) *Biosafety Projects*. www.unep.ch/biosafety. [Verified 20 April 2006].

6

Crops behaving badly: pollen spread, its prevention and coexistence of genetically modified crops with conventional varieties

The potential of pollen to spread from genetically modified (GM) crops to wild relatives is discussed. This will be compared to the spread of pollen from conventionally bred crops to their wild relatives, and the consequences of both types of spread. Means of preventing such spread and the overall question of the coexistence of GM crops with their conventional counterparts is reviewed.

Before considering the case of pollen movement from a GM to a non-GM crop it is useful to consider seed quality and purity in general. The maintenance of quality is vital to modern agriculture. International seed trade is one of the most regulated sectors of the agricultural economy. Quality is controlled by the Association of Official Seed Certification Agencies or the OECD Seed Certification system (Conner et al. 2003). The accidental presence of seed from another variety within a cultivar is called **adventitious seed**. Until fairly recently, seed purity was based on the **phenotype** of the plant. Now molecular genetic techniques enable genetic purity to be directly determined, but for both GM and non-GM plants, 100% purity is impossible in practice because of unintended mixing. In non-GM seed, this is recognised in the OECD system. It is also recognised that attainable levels of purity vary with the biological behaviour of the plant. For instance, it is easier to keep a strictly self-pollinating plant, such as wheat, free from other varieties than it is for a wind-pollinated crop such as maize. This is reflected in the recommended seed purity levels for certification.

Standards of purity vary in different countries. It may be as stringent as 100% or as relaxed as 95%. If the adventitious presence in a conventional crop is of an approved GM crop, it is between 0.9% and 5%. If the impurity is caused by a GM crop that is not approved for food or feed, it is 0% in almost all countries.

The first requirement for pollen spread from a GM crop is that it is planted near to a conventional variety of that crop or a wild relative that it can cross-pollinate. Most GM crops are not grown in areas where they could pollinate wild relatives. This applies to 87% of the world's soybean crop, and 95% of the world's maize and cotton. Even in those countries where cross-pollination could occur, such as in China for soybeans, wild relatives are found in only certain areas (Roush 2004). Cross-pollination is, however, possible with GM canola in Canada and the USA, and with maize in Mexico. The latter would be grown illegally as that country has a moratorium on the planting of GM maize although it has allowed the planting of GM cotton since 1998.

What harmful effects could result from such pollen spread? This is discussed in a book with an extremely 'sexy' title (Ellstrand 2003). This shows no evidence per se of harm to human health or to the environment, including no evidence of increased weediness due to pollen spread from GM crops. There is evidence that conventional agriculture with non-GM crops has adversely affected some wild plants through genetic swamping of their populations, and that some wild plants have become weedier during hybridisation with cultivated crop varieties.

According to Ellstrand (2003), the problems associated with hybridisation became a focus after the advent of transgenic crops. For example, hybridisation with cultivated rice has been implicated in the near extinction of an endemic Taiwanese wild rice. Indigenous cotton in the Galapagos Islands could be at risk of extinction or replacement as a result of hybridisation with cultivated cotton. Ellstrand provides similar evidence for at least another nine species. He also documents in great detail the history of sugar beets in Europe, where hybrids between cultivated beets and their progenitors, the sea beets, have caused major weed problems.

A recent example of the introgression of genes between wild and cultivated varieties of rice was recently published (Li et al. 2006). Domestication of rice and many other crops involved the selection of plants that did not naturally shed ripe fruits or seeds. In the case of rice, this was done by selecting the progeny of wild and domesticated rice.

In an assessment of the risks of transgene escape John Burke (2003) wrote that although it may take years for the full environmental effects of transgene escape to be known, predictions regarding the particular crop or traits that are likely to pose the greatest environmental risks can be made. Crops that hybridise readily with wild relatives may represent greater risks than those that do not. Similarly, transgenes that confer an advantage on potentially wild or weedy forms of a plant are likely to pose a significant risk. In contrast those that are neutral or disadvantageous are likely to have few negative impacts. Some of the concern at present stems from the fact that many of the transgenic traits – such as pest or pathogen resistance and tolerance of various abiotic stresses – may be highly advantageous in the wild.

In support of this Snow (2002) wrote that when novel genes spread to free-living plant populations they could create or exacerbate weed problems by providing novel traits that allow these plants to compete better, produce more seeds, and become more abundant.

This chapter aims to analyse what is known about gene flow from GM crops, their consequences and possible control.

Gene flow in oilseed rape, sunflowers, rice and potatoes

In a national assessment in the United Kingdom (UK) of hybridisation between cultivated, but non-GM oilseed rape, *Brassica napus*, and its wild counterpart, *B. rapa*, it was found that an estimated 32 000 hybrids form annually in seminatural waterside populations. This was somewhat less in agricultural weedy populations, which were found to contain about 17 000 hybrids during the same period (Wilkinson et al. 2003). The authors point out that, although hybrids occur, their presence is not a hazard in itself and does not imply inevitable ecological change. Hybrid fitness and other factors affecting the likelihood of such change needs to be assessed. However, what measures can be taken to prevent gene flow in cases where it is likely to occur?

Sunflowers (*Helianthus annuus*) are commonly afflicted by white mould (*Sclerotinia sclerotiorum*) infection, which begins at the base of the stem and results in the rapid wilting and death of plants, greatly reducing seed output. Infection rates as high as 100% have been reported in North America and white mould has been known to reduce yields by as much as 70%. Attempts to develop resistant cultivars via traditional plant breeding techniques have met with little success but GM-resistant varieties have been developed. The potential for transgene escape is especially high in sunflowers and crop–wild gene flow is almost a certainty throughout the range of sunflower cultivation in the USA (Burke 2003).

Results of experiments to test the effects of such gene flow showed that the transgene in cultivated varieties diffused neutrally, without any negative effects, into wild cultivars following its escape. It appeared that by giving the white mould resistance transgene to wild sunflowers they were effectively gaining something they already had, that is, some degree of mould resistance. These results show clearly the importance of a case-by-case analysis of the relative risks and benefits of genetic modification in the environment (Burke and Rieseberg 2003).

However, not all transgenes may be as neutral. In a study of the effects of a Bt gene transferred from cultivated to wild sunflowers, Snow et al. (2003) found that the resultant plants were less susceptible to insect pests and resulted in 55% more seed per plant.

Although no GM rice has been approved for commercial cultivation, genes conferring traits such as ß-carotene (vitamin A precursor), insect and disease resistance, and herbicide and salt tolerance, are in the pipeline. As rice is one of the four most important crops worldwide, the others being wheat, maize and potatoes, gene flow into weedy or wild relatives needs to be considered. Transgenes that are responsible for resistance to **biotic** and abiotic stresses can significantly enhance the ecological fitness of such relatives. Thus, understanding the likelihood of gene flow from rice to its weedy and wild relatives will help to predict the potential ecological consequences caused by transgene escape. This knowledge will also facilitate the effective management and safe use of transgenic rice. Studies have shown that such gene flow is indeed possible and is significantly higher when GM rice varieties are grown near to wild rice species (Lu 2004). However, the major wild species of concern is red rice, *Oriza rufipogon*, which is already a problematic weed and is unlikely to become any more of a nuisance than it already is in many rice-growing areas.

A genetically modified potato can be developed within five years rather than the eight to 15 years often needed for conventional breeding. The main centre of potato biodiversity is the central Andes. The question therefore arises as to whether GM potatoes should be grown in these regions. The Nuffield Council on Bioethics (2004) suggests that gene flow into related species in centres of crop biodiversity is not a sufficient justification to bar the use of GM crops in the developing world. They propose that not to allow this technology to be used in such regions in itself poses a socioeconomic risk to the poor. In the light of this, scientists studied gene flow from GM nematode-resistant potatoes to wild species in field trials in Peru (Cells et al. 2004). Nematodes are the principle pests of South American potato crops and resistance is conferred by a gene that encodes a proteinase inhibitor. This inhibits the action of the digestive enzymes, called proteinases, in the gut of the nematode, impairing digestion of its dietary protein.

Cells et al. (2004) found that gene flow did indeed occur. They then went on to study the effects of the male sterile potato cultivar Revolucion. This variety lacks viable pollen and is therefore unable to fertilise ovules of another compatible plant. Transgenic Revolucion potatoes were resistant to nematodes and were unable to cross-pollinate wild varieties. They therefore propose that transgenic plantings should be limited to male sterile cultivars while concerns are evaluated over several generations. However, many potato cultivars are female fertile, in which case introgression will happily occur by the transfer of pollen from the wild relatives into the GM crop. Hence, the variety of the potato will determine whether male sterility can inhibit gene flow and it would be advisable to use male sterile cultivars.

Gene flow from transgenic Bt maize to non-Bt maize refuges

The importance of the planting of non-Bt refuges, to prevent targeted insects developing resistance to the Bt toxin, was outlined in Chapter 2. Should pollen containing Bt genes from a GM crop fertilise the ovules of refuges the resulting seeds (kernels in the cob) could transfer the gene to the next generation. This could decrease the effectiveness of refuges as they will produce fewer susceptible insects than expected. In field studies designed to test this Chilcutt and Tabashnik (2004) planted the non-Bt refuges downwind of the Bt maize to maximise any chances of cross-pollination. They found that at one metre distance the maximum toxin concentration was 45% of the adjacent Bt plants. At 31 metres this level had dropped to 1.7%. They therefore suggest that the minimum width of non-Bt maize refuges be increased to at least 30 metres, from the current four metres as the EPA guidelines recommend. This is still under discussion by the EPA.

Gene flow in Mexican maize

Maize originated in Mexico and Central America, having been bred from the wild relative, teosinte. About two-thirds of the maize grown in Mexico consists of the local varieties known as landraces. These landraces have great cultural and humanitarian value (Snow 2005). Quist and Chapela (2001) gave rise to a worldwide debate. In it they claimed that they had found traces of transgenic DNA in native maize landraces growing in the remote mountains in Oaxaca, Mexico, part of the Mesoamercan centre of origin and diversification of maize. The publication of this paper immediately resulted in numerous scientists questioning the validity of the experimental techniques used by Qusit and Chapela. Finally the editor of *Nature* responded with the following statement:

> 'Nature *has concluded that the evidence available is not sufficient to justify the publication of the original paper'* (Nature 2002).

Despite this dispute, it is possible for seeds from GM maize grown in the USA to be carried across the border or arrive in the huge amounts of maize that Mexico imports from the USA each year. Once an intro-duction occurs the ability of the transgene to persist and multiply in the landraces depends on whether the resulting plants have a reproductive advantage. Farmers might also actively select transgenic plants when they save and trade seed (see Chapter 5 for possible effects on genetic diversity).

A coalition of Mexican farmers, environmentalists and representa-tives from indigenous communities asked for a study of the effects of GM corn on native maize and related plants such as teosinte. This chal-lenge was taken up by the Commission for Environmental Cooperation (CEC), an organisation established jointly by Canada, Mexico and the USA. Their report was made public in November 2004 (Fox 2005).

They called for a moratorium to be enforced on commercial plant-ing of GM corn in Mexico. They were, however, careful to say that GM corn is safe and useful in the USA and is 'probably even safe in Mexico'. Other recommendations in the report include the need to find better ways of conserving Mexican maize and teosinte races and to improve the coordination of biotechnology regulatory policies among Mexican, Canadian and US governments. Another suggestion is that further research should be undertaken into methods to monitor and mitigate gene flow. One of the major causes of the loss of teosinte has been the cultivation of conventional maize.

To answer some of these questions an extensive study was carried out during 2003 and 2004 to investigate the frequency of transgenes in landraces in Mexico (Ortiz-Garcia et al. 2005). They sampled maize seeds from 870 plants in 125 fields and 18 localities in the state of Oaxaca. They screened 153 746 seeds for the presence of two transgene elements. No such sequences were found and they concluded that 'transgenic maize seeds were absent or extremely rare in the sampled fields'.

During this debate it should be borne in mind that maize has lost the ability to survive in the wild due to its long process of domestication and needs human intervention to disseminate its seed. In addition maize is incapable of sustained reproduction outside of domestic cultivation and maize plants are non-invasive in natural habitats. Despite the fact that maize frequently appears as a volunteer plant in a subsequent crop rotation, it has no inherent ability to persist or propagate. In all regions of the world, volunteer plants are managed with herbicides, tillage or manual removal of plants.

Gene flow from genetically modified herbicide-tolerant crops

There is no doubt that transgenes will flow from GM crops to other plants with which they are compatible (i.e. which they can pollinate). If a gene encoding tolerance to a herbicide were to be transferred from a crop to a weedy relative, would it result in a 'superweed'? The term superweed is not a scientific one but one conjured up to generate fear of GM crops. But a superweed must be one that cannot be controlled. If a weed becomes resistant to just one herbicide such as glyphosate, however, it can be killed by another herbicide such as glufosinate.

A study was conducted in Canada involving gene flow among three different varieties of canola (Hall et al. 2000). Two of the varieties were transgenic and one was not. Each was resistant to one of three different herbicides, glufosinate, imidazolinone and glyphosate. Hybridisation occurred between all three varieties leading to a few **volunteers** that were resistant to all three herbicides. According to Hall the evolution of multiple resistance has not resulted in a major problem for most canola farmers but it has increased the complexity of weed control in the rotation (Hall et al. 2000). For instance, as most pollination of canola is by insects such as bees, isolation distances of 100 metres are normally sufficient for commodity oilseed production. However, regulations for the production of hybrid seed specify 800 metres.

Creeping bentgrass (*Agrostis stolonifera*) is a grass widely used on golf courses. Its species name reflects its ability to reproduce asexually by means of **stolons**. Seed dispersal by wind, water, wildlife and mechanical means may also occur. Commercial approval for bentgrass resistant to glyphosate is being sought from the United States Department of Agriculture. About 160 hectares of GM bentgrass were field tested in 2003, presenting a unique opportunity to test methods to track gene flow from the GM fields to compatible relatives in the surrounding, largely non-agronomic areas (Watrud 2005). The highest frequencies of gene flow from the GM crop were observed within about two kilometres in the direction of prevailing winds. However, pollen-mediated movement of the GM gene was recorded as far as 21 kilometres away. The overall frequency varied from 0.03% to 2% depending on the type of non-GM plant tested.

Coexistence of genetically modified and non-genetically modified crops

What is coexistence? According to the European Commission (2003), it relates to 'the economic consequences of (the unintended) presence of material from one crop in another and the principle that farmers should be able to cultivate freely the agricultural crop they choose'. In other words, coexistence should allow two different but related crops to be grown in proximity to each other without the adventitious presence of genes from the one being found in the other. Such presence could be caused by seed impurities, cross-pollination, volunteers, from seed planting equipment and practices, farm harvesting and storage practices, transport, storage and processing (Brookes 2003).

The presence of GM traits in non-GM crops has primarily become an issue because of the development of distinct markets for non-GM derived products. Initial demands from some consumers were for zero tolerance of GM material. This was a unique requirement as, although not widely known among consumers, no food products traded or consumed anywhere is required to be 100% pure. All traded products operate under conditions that recognise tolerances for the presence of some other material. For example, in organic agriculture there is a tolerance for the use of 5% of some non-organic ingredients in processed products. The burden of any additional costs incurred in meeting tolerances for the adventitious presence of unwanted material has, to date, invariably fallen on the product or sector that seeks to gain a market or economic advantage from the crop subject to integrity preservation such as organic produce (Brookes 2003).

Another example of tolerance of unwanted material is in the case of certified seed production in which seeds are classified according to various purity levels. These are based on specified seed separation distances and time intervals between the seed crop and any other crop of the same species grown on a plot. Failure to meet the purity standards results in seed not being certified and the relevant seed premium is lost to the grower. In other words, the crops have to be sold as a non-seed crop. Compliance with these standards shows that in more than 96% of cases the procedures adopted, including isolation, cleaning, rotations and separation of harvest, are sufficient to meet the stringent purity standards (Brookes 2003).

An extremely important case of coexistence is in the production of varieties of oilseed rape for either industrial use or human consumption as contamination of the latter by the former could lead to human deaths. High concentration of the fatty acid, erucic acid, is important for industrial oils but is toxic to humans. In contrast, varieties containing negligible erucic acid are used in animal feed and human food, one example of the latter being canola. The threshold for mixture of high erucic oil seeds in other rape is 2%. Evidence from Europe suggests that a 100-metre separation distance tends to result in more than 95% of fields of non-erucic plants having an adventitious presence of high-erucic plants at the level of 0.2% to 0.5%. Adherence to the separation distances comes by voluntary arrangements between adjacent farmers, although in many instances farmers producing both types of oilseed rape handle this separation themselves (Brookes 2003).

A study in Australia showed that pollen flowed up to 2.6 kilometres from GM herbicide-tolerant varieties. Despite this relatively long travel distance, the highest frequency of cross-pollination measured within this radius was 0.23% with 69% of samples tested showing no outcrossing (Rieger et al. 2002). In addition it was found that herbicide-tolerant rape volunteers are not a significant problem and, where they have occurred, farmers have used a variety of herbicides to control them. Finally, while there is a possibility of gene transfer between different cultivated rape species, the potential for hybridisation in the field, introgression of a herbicide-tolerant trait and its subsequent stable expression in the progeny of weeds, is considered to be a low and manageable risk (Brookes 2003).

The evidence shows that GM crops, at least in the USA, have coexisted with conventional and organic crops without significant economic or commercial problems (Brookes and Barfoot 2004b). Surveys among US organic farmers shows that most (92%) have not incurred any direct, additional costs or incurred losses as a result of GM crops having been grown near their crops. However, some Canadian organic farmers claim that they cannot produce organic canola as a result of market concerns about cross-pollination with GM canola. Although official figures of the extent of Canadian organic canola plantings are not available, trade sources estimate that this amounted to about 2000 hectares out of a total of 4.69 million hectares, equivalent to 0.04% of the crop.

Some analysts have suggested that the lack of coexistence conditions for growing herbicide-tolerant GM canola has resulted in problems for both GM and non-GM canola farmers (see Brookes and Barfoot 2004b). This is largely as a result of lack of control of volunteers that are herbicide tolerant. This is not a specific GM problem as it relates to non-GM herbicide tolerance traits as well. Indeed the Canola Council of Canada (2001) reported that 60% of farmers found that volunteer management was the same as before the advent of GM canola, 16% indicated it was easier and 23% thought it was more difficult. This suggests that volunteers do not appear to be a problem for most of the farmers who took part in the survey.

Several organic certification agencies in North America offer advice on ways of ensuring product integrity. These include:

- seed verification;
- site selection with respect to wind direction to minimise cross-pollination;
- good neighbour relations with farmers growing GM crops;
- equipment cleanliness;
- harvest testing for the adventitious presence of GM traits;
- storage inspection;
- transportation management; and
- meticulous record keeping.

Has the growth of GM crops impeded the development of organic crops? In the USA, states with significant levels of organic maize and soybean are often those in which high percentages of GM varieties are also planted. This suggests that there have been few coexistence problems. The court action taken by a group of organic canola farmers in Saskatchewan against the providers of herbicide-tolerant GM canola implies that at least in that province coexistence might be a problem. The passage of this action may provide an opportunity to assess this further.

Genetically modified crop coexistence and economic consequences

There are two main economic implications for GM and non-GM coexistence:

1 the costs involved in complying with the limitations on the presence of unwanted material (tolerance thresholds); this currently refers to GM in non-GM products – could this in future include the presence of non-GM in GM products? and

2 the economic consequences of not meeting limitations that could include the loss of non-GM or organic price premiums.

Costs involved in meeting tolerance thresholds

By and large the tighter the threshold (the lower the limitation), the higher the cost involved in meeting that tolerance. This is, true for all crops, GM or non-GM. However, in the GM field in the European Union (EU) market during 2002/03, non-GM market premiums for soybeans and soymeal varied from 2% to 5% for a 1% tolerance of GM material. This rose to 7% to 10% for a 0.1% tolerance. These levels, however, relate to imports, not GM crops grown in Europe.

Work in the UK has estimated the separation distances required to meet various tolerance levels for sugar beet, maize and oilseed rape (Table 6.1). These, however, are non-seed crops, grown as commodities, not for the production of seed.

Table 6.1 Recommended separation distances to ensure cross-pollination is below specified limits in non-seed crops of sugar beet, maize and oilseed rape

	Threshold levels of cross-pollination (m)		
	1%	0.5%	0.1%
Oilseed rape	1.5	10	100
Maize grain	200	300	Insufficient information
Sugar beet[A]	0	0	0

From Brookes (2003)

[A] There are no recommended separation distances for sugar beet as the crop is usually harvested before flowering.

Oilseed rape is often cited as the problem crop with regard to coexistence. To investigate this, studies in Australia showed that pollen flowed up to 2.6 km from GM herbicide-tolerant varieties. Despite this relatively long travel distance, the highest frequency of cross-pollination measured within this radius was 0.23% with 69% of samples tested showing no outcrossing (Rieger et al. 2002). In addition it was found that herbicide-tolerant rape volunteers are not a significant problem and, where they have occurred, farmers have used a variety of herbicides to control them. Finally, while there is a possibility of gene transfer between different cultivated rape species, the potential for hybridisation in the field, introgression of a herbicide-tolerant trait and its subsequent stable expression in the progeny of weeds, is considered to be a low and manageable risk (Brookes 2003).

In the UK Farm Scale Evaluations of herbicide-tolerant crops (see Chapter 3), coexistence cost issues include costs incurred by farmers to comply with conditions for growing GM crops, and costs incurred by farmers growing non-GM crops near the GM evaluation farms. In the former cases, 60% of the growers indicated that the procedures required were in line with other farm assurance schemes. The only new costs involved the audit requirements of the UK government at £800 (pounds sterling) per site (area of GM crops planted on a farm). This could give some indication of what farmers growing GM crops commercially might encounter, although with wider competition these costs might reasonably be expected to fall to levels in line with other quality assurance schemes in the order of £0.44 to £1.4 (sterling) per hectare (Brookes 2003).

The main focus of attention relating to possible coexistence costs incurred by non-GM farmers who are near the evaluation farms are related to organic farming. Organic accreditation bodies recommend a distance of up to six km depending on wind direction, time of flowering and local topography. However, this is the maximum, and in practice, depending on the crop, distances vary from 500 m to 6000 m. In most cases where organic farming occurred within six km of an evaluation farm, most were not classified at risk (Brookes 2003).

The economic consequences of exceeding the thresholds for adventitious presence of genetically modified organisms

The incentive for any non-GM producing farmer, including those who grow both GM and conventional varieties, to undertake measures aimed at minimising adventitious presence depends on the relative costs involved compared to the economic consequences. Such consequences might include the loss on non-GM price advantages or the inability to sell the non-GM crop in a given market. An example of the lack of concern about such consequences can be seen in the case of Spanish farmers growing GM maize. As GM and non-GM maize trade at the same price there was no incentive for farmers to keep the varieties separate.

To examine the Spanish example, out of about 460 000 hectares planted to maize, about 12% in 2004 was insect-resistant GM, 0.2% organic and the majority conventional. The evidence shows that these three types of maize production have coexisted without economic or commercial problems. This includes regions where GM maize is concentrated, accounting for 15% of total maize plantings (Brookes and Barfoot 2003).

Coexistence and liability

When farmers consider switching from one form of agriculture to another, whether it is from non-GM to GM or from conventional to organic, they will weigh up the costs of such conversion. This could include the impact on yield, costs of production, compliance with standards and these must be compared with the relative benefits of, for example, higher prices. The higher the compliance costs, the lower would be the incentive to switch, and vice versa. In most agricultural markets, the burden of costs associated with maintaining the integrity of a product, or 'preserving its identity' falls on the sector that produces that product and is seeking to benefit from its production.

If regulation of GM crops imposed liability on growers for possible impact on non-GM varieties, this would set a precedent in farming regulations. Any such precedent should, however, have to apply to all farmers including non-GM and organic crop producers, whose activities might have an adverse impact on GM crop producers (Brookes 2003).

Preventing and managing gene flow
Targeting genes to chloroplasts

Chloroplasts are small organelles found in plants. They contain the green pigment, chlorophyll, have their own chromosome and are responsible for providing the plant with energy derived from the sun in the process known as photosynthesis. They are mostly inherited maternally and are hardly ever transferred by pollen. Targeting transgenes to the chloroplast, therefore, will in most cases overcome the problem of gene flow.

In addition, a transgene can be targeted to a specific region of the chloroplast genome, making sure it does not disrupt important genetic functions of the plant. This is not yet possible when transgenes are integrated into the nuclear genome. Another major advantage is that chloroplasts are present in up to 100 copies in plant cells. Thus, targeting a transgene to chloroplasts immediately results in increased production of the protein coded for by that gene.

In one example of overexpression of the Bt gene in tobacco chloroplasts such high levels were obtained that the toxin molecules formed into crystals in much the same way they do in the *Bacillus thuringiensis* bacterium (De Cosa et al. 2001). In another example the overexpression in the chloroplast conferred resistance even to normally Bt-resistant insects (Kota et al. 1999).

Unfortunately, this approach can only succeed if cross-pollination is from the male to the female parent. In many cases cross-pollination goes in both directions and hybrids will also be produced when the GM crop is the female parent. So far this technology has not been applied to agricultural crops. However, it could be extremely useful in the production of pharmaceuticals in plants (see Chapter 9).

Seed sterility

An example of the use of seed sterility is the system dubbed 'terminator technology'. It is a way of making GM plants produce sterile seeds. This was patented by the United States Department of Agriculture together with university-based scientists. It was licensed to Delta & Pine Land Seed Company. To date it has not been commercialized. The technology is not unlike that used by the entertainment and publishing industries which strive to protect their investments from the illegal copying of films, music and books.

Critics say the technology was developed to force farmers to buy fresh seed every year instead of saving it from the previous year's harvest. However, many farmers, even poor ones in developing countries, buy seed every year (see Table 2.5). This is particularly so with hybrid seed in crops such as maize where farmers know that if they plant their own seed, their crops will lose the advantages of hybrid vigour.

Another example is with cotton, an in-bred crop for which farmer-saved seed would, in principle, be possible. However, even in developing countries, farmers still buy cotton seed of varieties determined by the cotton buyers. This is so that the farmer produces a crop that meets the quality requirements of their customer. In addition, no farmer is forced to buy GM crops. Farmers are 'savvy' – they will only plant GM crops if they improve yield, cut import costs or increase their profits.

However, there is another reason why the anti GM-activists have misunderstood this issue. Anyone concerned about the spread of pollen from GM crops has good reason to encourage the use of the 'terminator technology'. Certainly, in GM crops that have enhanced survival properties, this technology would prevent cross-pollination with weedy or other unwanted relatives and thus stop gene flow. The term 'terminator technology' is emotive. The current term is gene use restriction technology (GURT) but it is doubtful that it will replace the original term.

A series of 'natural terminator technologies' have been used long before this term was coined. These include seedless melons, grapes and other crops that have been cultivated with the express purpose to avoid having seed. The easiest way to breed such varieties is to make the crop 'triploid' (i.e. it carries three sets of chromosomes). Triploid plants produce fruit but the seeds all abort.

Isolation distances

An obvious strategy is to make sure that GM crops are planted far enough away from wild relatives to make cross-pollination unlikely. These distances, aimed at achieving 99% or greater levels of genetic purity, had been developed for many crops long before GM plants were grown. Distances can range from hundreds of metres for maize, to over 1000 metres for sugar beets. For crops such as wheat that are primarily self-fertilising, isolation distances can be as low as zero. However, these isolation distances were established to prevent wild pollen flowing into the crop. What is being attempted here is to stop outflow of pollen from a GM field.

Growing plants that interfere with pollen flow

Planting what is knows as 'border rows' or 'pollen trap crops' are often used to enhance the effects of isolation distances. These can either be plants of a different species or a different variety of the same species. For example, sugar beet breeders plant a strip of non-drug producing hemp, *Cannabis sativa*, around their seed multiplication plots. These relatively tall plants have sticky leaves that trap incoming pollen from other cultivars (Ellstrand 2003). Similar traps are used with GM cotton and maize (see Chapter 2).

Summary

Gene flow from GM crops is inevitable. Therefore, the consequences need to be understood. Most studies have shown that gene flow will have a fairly limited economic impact on non-GM crops and organic farming.

References

Biocom Ag (2006) *Information and Communication for Biotechnology.* http://www.biocom.de [Verified 5 June 2006.]

Brookes G (2003) Co-existence of GM and non GM crops, economic and market perspectives. (PG Economics: Dorchester, UK.)

Brookes G and Barfoot P (2003) Co-existence of GM and non GM arable crops, case study of the UK. (PG Economics: Dorchester, UK.) http://www.pgeconomics.co.uk [Verified 30 April 2004].

Brookes G and Barfoot P (2004a) Co-existence of GM and non GM crops, case study of maize grown in Spain. (PG Economics: Dorchester, UK.) http://www.pgeconomics.co.uk [Verified 30 April 2004].

Brookes G and Barfoot P (2004b) Co-existence in North American agriculture, can GM crops be grown with conventional and organic crops? (PG Economics: Dorchester, UK.) http://www.pgeconomics.co.uk [Verified 30 April 2004].

Brookes G, Barfoot P, Melé E, Messeguer J, Bénétrix F, Bloc D, Foueillassar X, Fabié A and Poeydomenge C (2004) Genetically modified maize, pollen movement and crop co-existence (PG Economics: Dorchester, UK.) http://www.pgeconomics.co.uk [Verified 30 April 2004].

Burke JM (2003) Assessing the risks of transgene escape, a case study in sunflowers. Information Systems for Biotechnology (ISB) News Report Sept. pages 2–4.

Burke JM and Rieseberg LH (2003) Fitness effects of transgenic disease resistance in sunflowers. *Science* **300**, 1250.

Canola Council of Canada (2001) An agronomic and economic assessment of transgenic canola, Canola Council, Canada. www.canola-council.org [Verified 20 June 2006].

Cells C, Scurrah M, Cowgill S, Chumblauca S, Green J, Franco J, Main G, Klezebrink D, Visser RGF and Atkinson HJ (2004) Environmental biosafety and transgenic potato in a centre of diversity for this crop. *Nature* **432**, 222–225.

Chilcutt CF and Tabashnik BE (2004) Contamination of refuges by *Bacillus thuringiensis* toxin genes from transgenic corn. *Proceedings of the National Academy of Sciences of the USA* **101**, 7526–7529.

Conner AJ, Glare TR and Nap J-P (2003) The release of genetically modified crops into the environment Part II. Overview of ecological risk assessment. *The Plant Journal* **33**, 19–46.

De Cosa B, Moar W, Lee S-B, Miller M and Daniell H (2001) Overexpression of the Bt Cry2Aa2 operon in chloroplasts leads to formation of insecticidal crystals. *Nature Biotechnology* **19**, 1–4.

Ellstrand NC (2003) *Dangerous Liaisons, When Cultivated Plants Mate with their Wild Relatives*, p. 195. (Johns Hopkins University Press: Baltimore, USA.)

European Commission (2003) *Communication on Co-existence of Genetically Modified, Conventional and Organic Crops*, C March 2003.

Fox JL (2005) Report recommends ban of US GM maize in Mexico. *Nature* **23**, 6.

Hall L, Topinka K, Huffman J, Davis L and Allen A (2000) Pollen flow between herbicide-resistant *Brassica napus* is the cause of multiple-resistant *B. napus* volunteers. *Weed Science* **48**, 688–694.

Kota M, Daniell H, Varma S, Garczynski SF, Gould F and Moar WJ (1999) Overexpression of the *Bacillus thuringiensis* (Bt) Cry2Aa2 protein in chloroplasts confers resistance to plants against susceptible and Bt-resistant insects. *Proceedings of the National Academy of Sciences of the USA* **96**, 1840–1845.

Li C, Zhou A and Sang T (2006) Rice domestication by reducing shattering. *Science* **311**, 1936–1939.

Lu B-R (2004) Gene flow from cultivated rice, ecological consequences. *Information Systems for Biotechnology News Report* May 2004, 4–6.

Nature (2002) Editorial note. *Nature* **416**, 600.

Nuffield Council on Bioethics (2004) *The Use of Genetically Modified Crops in Developing Countries*. (Nuffield Council on Bioethics: London.)

Ortiz-Garcia S, Ezcurra E, Schoel B, Acevedo F, Soberón and Snow AA (2005) Absence of detectable transgenes in local landraces of maize in Oaxaca, Mexico (2003–2004). *Proceedings of the National Academy of Sciences of the USA* **102**, 12338–12343.

Quist D and Chapela IH (2001) Transgenic DNA introgressed into traditional maize landraces in Oaxaca, Mexico. *Nature* **414**, 541–543.

Rieger M, Lamond M, Preston C, Powles SB and Roush RT (2002) Pollen-mediated movement of herbicide resistance between canola fields. *Science* **296**, 2386–2388.

Roush R (2004) Crops behaving badly. *Nature* **427**, 395–396.

Snow AA (2002) Transgenic crops – why gene flow matters. *Nature Biotechnology* **20**, 542.

Snow AA (2005) Genetic modification and gene flow, an overview. In *Controversies in Science and Technology, from Maize to Menopause*. (Eds DL Kleinman, AJ Kinchy and J Handlesman.) pp. 107–118. (University of Wisconsin Press: Madison, WI.)

Snow AA, Pilson D, Rieseberg LH, Paulsen MJ, Pleskac N, Reagon MR, Wolf DE and Selbo SM (2003) A Bt transgene reduces herbivory and enhances fecundity in wild sunflowers. *Ecological Applications* **13**, 279–286.

Watrud LS (2005) Long distance pollen-mediated gene flow from creeping bentgrass. *Information Systems in Biotechnology* **January 2005**, 1–3.

Wilkinson MJ, Elliott LJ, Allainguillaume J, Shaw MW, Norris C, Welters R, Alexander M, Sweet J, Mason DC (2003) Hybridization between *Brassica napus* and *B. rapa* on a national scale in the United Kingdom. *Science* **302**, 457–459.

7

When plants' genes don't come from their parents – horizontal gene transfer

What is horizontal gene transfer? The normal reproductive transfer of genes from parents to offspring is regarded as vertical gene transfer. Gene transfer from one species to another is referred to as horizontal gene transfer (HGT). In this chapter the horizontal transfer of genes from transgenic plants to other organisms is discussed. These genes could be to bacteria in the intestine, after human or animal ingestion of transgenic plant material, or to organisms in the soil after release of plant transgenes into the soil. The likelihood of such events occurring and the consequences of such horizontal gene transfer is considered. From the recent sequencing of the genomes of many organisms it is clear that HGT has been occurring naturally over millennia. Indeed it has been estimated that up to 16% of bacterial genes have been acquired via HGT (Ochman et al. 2000).

In order for genes to be transferred from a GM plant to any other organism the following steps are required: availability of the genetic material for transfer, an uptake mechanism by the other organism and a method for the establishment of the transferred DNA in that organism.

In addition, if the transferred DNA is to have any effect on the recipient organism, the fragment of DNA must be expressed so as to confer a new trait on that organism (Conner et al. 2003).

Availability of DNA

At some stage during the growth, death and decay of a GM plant its DNA should be available for transfer. The DNA would need to persist in the environment for long enough periods to allow uptake to occur. In addition, in order to confer a new trait on the recipient organism the fragment of DNA would also have to be large enough to code for a protein.

Uptake of DNA

The recipient organism must have a mechanism for the uptake of the DNA released from the GM plant. There are several ways in which bacteria can take up DNA.

- Transformation. This is the method whereby bacteria take up free DNA from their surroundings. In order to do this bacteria have to become 'competent' – a process in which the bacteria produce the necessary proteins and enzymes required to take up free DNA. Over 40 bacterial species have been shown to be naturally competent for transformation. However this is likely to be an underestimate resulting from the inability to cultivate many bacterial species in the laboratory (Nielsen 2003). This is probably the most likely way in which bacteria could take up plant DNA.

- Transduction. This is the transfer of DNA to bacteria via a bacterial virus. These viruses have a limited host range and none has been found that can infect both plants and bacteria. Thus this method is highly unlikely to mediate HGT from plants to bacteria.

- Conjugation. Again this is a method for the transfer of genes between bacteria. It is the closest that bacteria get to sexual reproduction in that a donor bacterium forms a conjugation bridge to a recipient and transfers some of its genetic material via this bridge. Thus, the donor could be thought of as a 'male' bacterium and the recipient as a 'female'. There is a species of bacterium that is able to transfer some of its DNA into plants by a method similar to conjugation. It is therefore conceivable that the reverse transfer could occur but it has never been demonstrated in the laboratory.

Establishment and expression of the transferred DNA in the recipient

The third requirement for HGT is the stable establishment of the transferred DNA in the genome of the recipient. The recipient needs to have a mechanism to incorporate and maintain the incoming DNA. In addition, for the transferred DNA to confer a new trait to the recipient, the DNA must successfully encode a protein. In order for the transferred DNA to be maintained and stably expressed in the recipient the new trait should not impose any negative effects on the recipient.

How feasible is it that DNA from GM plants could satisfy the above conditions? In the case of GM plant material fed to animals such as ruminants, no reports exist of bacteria being able to take up DNA in this environment (Chrispeels and Sadava 2003). Similar lack of evidence is found in humans (Nielsen and Townsend 2004).

The plant environment offers a highly diverse habitat for microorganisms especially within the rhizosphere (the area immediately adjacent to the roots), the phyllosphere (the surface of leaves and other plant tissue) and even within plant tissues. Many of these bacteria could be exposed to DNA from lysed plant material but to date there is no evidence of the stable transfer of such genetic material.

Several studies have been conducted to monitor HGT during field trials of GM crops or in the laboratory using simulated field conditions. Some types of soil bacteria can take up fragments of plant transgenes under highly optimised conditions (e.g. De Vries et al. 2001; Kay et al. 2002). However, the field trials have not yielded evidence for horizontal gene transfer under natural field conditions (Gebhard and Smalla 1999).

More recently, however, scientists have questioned the ability of methods used in field studies to detect extremely low levels of horizontal gene transfer (Heinemann and Traavik 2004; Nielsen and Townsend 2004). They argue that current methods of environmental sampling are not sensitive enough. In effect such transfer could be occurring approximately a trillion times lower than the current risk assessment literature estimates it to be. They call for new and improved methods to be developed (Heinemann and Traavik 2004).

Similar criticisms have been made about studies of HGT from plant transgenes to bacteria in the intestine (Netherwood et al. 2004). The authors of the study concede that 'current methodology may not have detected the transformants if transfer were an extremely rare event'.

Most authors agree that horizontal gene transfer may occur but at extremely low frequencies. If that is so, it must be assumed that over time and space it will occur. Therefore, the consequences of such transfer in the intestines of humans and animals, and in the soil, are considered here.

These issues cannot be generalised and need to be assessed on a case-by-case and gene-by-gene basis. Most concerns have focused on the transfer of antibiotic resistance from a plant to a bacterium. During the generation of transgenic plants antibiotic resistance markers have been used during the selection process. Although these markers are being phased out, some transgenic plants in commercial use do contain these genes. Fears have been expressed that such antibiotic resistance genes could exacerbate the already alarmingly high levels of antibiotic resistance in pathogenic bacteria. The antibiotic resistance gene most widely used in the development of GM plants codes for kanamycin resistance. There is already widespread resistance to kanamycin in intestinal and soil bacteria. In addition, HGT between bacteria is common. Therefore, bacteria are far more likely to receive an antibiotic resistance gene such as kanamycin resistance from another bacterium than from a GM plant (Syvanen 2002).

A similar situation is found with soil and plant-associated bacteria being exposed to GM plants carrying the Bt toxin gene (see Chapter 2). This gene originates from a soil bacterium, *Bacillus thuringiensis*.

Therefore, bacteria in the environment are far more likely to obtain this gene from strains of *B. thuringiensis* in their surroundings than from GM plants. In addition, the Bt toxin gene introduced into plants has been changed in order for it to be efficiently expressed in plants instead of bacteria. This will tend to reduce its effectiveness if, by some very rare event, it were to be transferred into bacteria.

The same is likely to be the case with herbicide-resistant GM plants (see Chapter 3). Most of the genes involved have been derived from soil bacteria that are widely distributed in nature.

As increasing numbers of genomes are sequenced, it seems that many organisms are mosaics of other organisms. They have presumably received these foreign genes by mechanisms promoting HGT. As this has clearly been going on for millennia, it is highly unlikely that the transfer of a transgene from a GM plant would pose a threat to the environment.

Summary

Horizontal gene transfer from GM plants to bacteria, either in the plant environment or in the intestinal tract of animals and humans ingesting them, depends on several factors. These include the availability of the DNA, mechanisms for its uptake, and establishment and expression of the transferred DNA in the recipient organism. Studies have shown that this is highly unlikely. In addition, as most of the genes that have been used in the development of GM plants have been derived from soil bacteria, the likelihood of their transfer from such bacteria is much more likely than their transfer from GM plants.

References

Chrispeels MJ and Sadava DE (2003) *Plants, Genes and Crop Biotechnology*. (Jones and Bartlett: Sudbury, Massachusetts, USA.)

Conner AJ, Glare TR and Nap J-P (2003) The release of genetically modified crops into the environment. Part II. Overview of ecological risk assessment. *The Plant Journal* **33**, 19–46.

De Vries J, Meier P and Wackernagel W (2001) The natural transformation of the soil bacterium *Pseudomonas stutzeri* and *Acinetobacter* sp. by transgenic plant DNA strictly depends on homologous sequences in the recipient cells. *FEMS Microbiology Letters* **195**, 211–215.

Gebhard F and Smalla K (1999) Monitoring field releases of genetically modified sugar beets for persistence of transgenic plant DNA and horizontal gene transfer. *FEMS Microbiological Ecology* **28**, 261–272.

Heinemann JA and Traavik T (2004) Problems in monitoring horizontal gene transfer in field trials of transgenic plants. *Nature Biotechnology* **22**, 1105–1109.

Kay E, Vogel TM, Bertolla F, Nalin R and Simonet P (2002) In situ transfer of antibiotic resistance genes from transgenic (transplastomic) tobacco plants to bacteria. *Applied and Environmental Microbiology* **68**, 3345–3351.

Netherwood T, Martin-Orue SM, O'Donnell AG, Gockling S, Graham J, Mathers JC and Gilbert HJ (2004) Assessing the survival of transgenic plant DNA in the human gastrointestinal tract. *Nature Biotechnology* **22**, 204–209.

Nielsen KM (2003) An assessment of factors affecting the likelihood of horizontal gene transfer of recombinant plant DNA to bacterial recipients in the soil and phytosphere. In *Collection of Biosafety Reviews*. International Centre for Genetic Engineering and Biotechnology, Trieste, Italy. (ICGEB: Trieste.)

Nielsen KM and Townsend JP (2004) Monitoring and modelling horizontal gene transfer. *Nature Biotechnology* **22**, 1110–1114.

Ochman H, Lawrence JG and Groisman EA (2000) Lateral gene transfer and the nature of bacterial innovation. *Nature* **405**, 299–304.

Syvanen M (2002) Chapter title?. In *Horizontal Gene Transfer*, 2nd edn. (Eds M Syvanen and CI Kado.) pp. 237–239. (Academic Press: San Diego.)

Thomson JA (2002) *Genes for Africa. Genetically Modified Crops in the Developing World*. (UCT Press, Cape Town, South Africa.)

8

Biosafety regulatory, trade and legal issues

Regulating transgenic crops

To reap the many potential benefits from genetically modified (GM) crops, they must be safe to humans and the environment. Some countries, such as the USA and most other industrialised countries, have had regulatory processes for these crops in place for more than 10 years. Some agencies require extremely detailed information. For instance in the UK more than 40 questions have to be answered. These include (Halford 2003):

- the sexual compatibility of the plant with other cultivated or wild plant species;

- the geographical distribution of the species;

- the size and function of any regions of the host plant genome that have been deleted as a result of the modification;

- information on when and where in the plant the novel gene(s) is/are active and the methods used for determining this;

- any potentially significant interactions with non-target organisms;

- the location and size of the release site(s);

- the method for preparing and managing the release site(s), prior to, during and after the release, including cultivation practices and harvesting methods;

- a description of post-release treatment methods for the GM plant material;

- the likelihood and consequences of theft of GM material from the trial, and other risks such as vandalism of the trial and movement of modified material from the trial site in field machinery.

Other countries are in the process of establishing their regulatory processes to implement their commitments under the Cartagena Protocol, which sets out the ground rules that permit international trade in and planting of GM crops. What are needed are strong, but not stifling, regulatory systems for GM crops. As more is learned about transgenic crops and how to control any potential harmful effects, the need for regulation may decrease. However, until that happens, regulations are needed to provide the public with the confidence that this new technology is not harmful to them or to the environment.

Food safety is also assessed and products regulated. Although this book is about environmental impacts of GM crops, a short analysis of the status of the safety assessments of foods derived from such crops should be considered.

Two studies on the comparison of proteins in a variety of non-GM and GM potatoes have been carried out. Lehesranta et al. (2005) found that the natural variation observed in the non-GM samples was considerably greater than the differences between non-GM and GM varieties. Statistical analysis showed no clear differences between the protein patterns of the GM lines and their controls.

In the second study, Catchpole et al. (2005) compared the composition of field-grown GM and conventional varieties of potatoes. They found that the former were substantially equivalent to traditional cultivars. Indeed a major finding was the large variation among conventional cultivars. The changes brought about by conventional breeding were at least of a comparable magnitude to those resulting from genetic engineering.

Finally on the issue of food safety, a review entitled 'Food Safety Evaluation of Crops Produced through Biotechnology' (Chassy 2002) concluded that after extensive testing and marketplace experience, there is no need to question the safety of current transgenic crops.

Moving on to biosafety regulations, Greg Jaffe of the Centre for Science in the Public Interest in the USA has analysed the regulatory processes in five countries (Jaffe 2004). His findings are summarised as follows.

The USA

In the USA the United States Department of Agriculture (USDA) regulates the import, transport and release of GM crops. Following field testing, a petition for non-regulated status may be submitted to the USDA who conducts an environmental assessment and seeks public input. Until 2004 the USDA has deregulated 60 GM **events**. The crop species for which one or more events have been deregulated are maize, cotton, oilseed rape (canola), soybean, tomato, potato, squash, papaya, chicory, sugar beet, rice and flax.

The United States Environmental Protection Agency (EPA) regulates the release of GM crops that express a pesticide such as the Bt toxin. To register a toxin, the EPA must determine that it is unlikely to cause unreasonable adverse effects on the environment or to humans.

The United States Food and Drug Administration (FDA) regulates the food safety of all GM crops except the toxins mentioned above. The FDA has determined that GM crops are not food additives requiring pre-market approval.

Jaffe (2004) comments that there are very few data on the time and costs involved in the regulatory approval of GM crops in the USA. As this has important implications for approval of crops produced in the developing world, such information would be invaluable. Health traits would include lack of toxicity or allergenicity when tested in animals. Environmental traits would include lack of pollen spread to potentially weedy relatives, lack of horizontal transfer of genes to soil organisms and negative impacts on biodiversity. The cost could vary from US$1 million to US$5 million.

The European Union

To release a GM organism into the environment for field testing the company or individual must submit a notification to the competent authority of the member state where the release will take place. Sufficient information must be given to allow an environmental risk assessment to be performed. The member state sends a summary of the applications submitted to them to the European Commission, which forwards it to the other member states. These may ask for the full dossier and may give observations, but do not have to consult their public about experimental releases in other countries, and cannot stop a field trial in another member state.

If a GM organism is intended to be marketed, a more elaborate approval process is required. In this case the notification is again sent to one member state and must contain information of the potential impacts on human health and the environment, as well as a monitoring plan and proposal for labelling and packaging. The competent authority of the reporting country produces a report indicating whether or not the GM organism should be marketed. Again the report is sent to the Commission and other member states for comment. The difference is that now the application file goes through a complex process of scientific review by an EU-wide panel of scientific advisors, followed by a proposal for decision by a committee of the National Competent Authorities of the member states. At the end of the process the competent authority of the country where the application file was originally submitted provides consent for marketing the GM organism for no more than 10 years.

There are no data on how long it will take a transgenic crop to be approved by the EU since there has been a moratorium on new approvals since 1998. With the passage of the EU's new labelling and traceability requirements in 2003, it was anticipated that approvals would begin again in 2004. However, this has not happened. This is because the EU has restarted the process of approvals for import of GM crops for food and feed use as a result of a complaint to the World Trade Organization (WTO) by the USA, Canada and Argentina (see below), but has not brought any dossier to completion for large-scale planting of GM crops. Even these food/feed approvals are contested, as they have all been obtained after a procedural battle, and several member states refuse to apply them.

South Africa (environmental release only)
In 1997 South Africa enacted the Genetically Modified Organisms Act that requires a risk assessment to be submitted to obtain a permit from the Minister of Agriculture for the release of a GM organism (trial or commercial). In 1999 regulations pertaining to the Act were issued which, among others, requires the applicant to notify the public of any proposed release. In addition the regulations allow the consideration of the socioeconomic impact that the GM organism may have on the community living near the release site.

Taiwan (food safety only)
The Department of Health requires all GM foods to be registered. An applicant must conduct a food safety assessment that is evaluated by an advisory committee composed of government and non-government scientists.

Argentina
To conduct a field trial a developer submits an application that includes information about the GM organism and possible environmental effects. If this is approved, a license is issued and inspectors visit at least once to verify compliance with the license conditions. If a developer wants to plant a large-scale plot for non-commercial release, such as seed production, the applicant can request that the crop be 'flexi-

bilised'. This requires detailed risk assessments. If a permit is issued, isolation distances and confinement are no longer required. Once a crop is flexibilised no additional environmental review is needed for commercialisation.

A strong regulatory system

Jaffe (2004) considers what constitutes a strong and effective regulatory system. This must protect human health in that only products that are safe to eat will be marketed. The environment must also be protected as the risks for each transgenic crop must be thoroughly analysed before being released and any risks must be minimised or managed before release. If such a system is in place, consumers should trust the regulatory process.

Pre-market approval

Several countries have established new laws and regulations to cover transgenic crops. In the EU, South Africa and Argentina, crops are subject to food safety and environmental approval processes before the food is marketed or the crop released. In the USA, by law, developers are not formally obliged to submit an application to the FDA. In reality there is no case of a company using that loophole, because the consequences of any health allegation against the product in such a case would be devastating for the developer. Every GM crop grown on a large scale has to go through the USDA approval system.

Transparency and public participation

As much information as possible should be supplied to the public regarding the regulatory process. This will exclude genuine confidential business information. This is certainly the case in the pharmaceutical industry and the food industry for non-GM innovations. The reason why even aspects of safety testing data are claimed as confidential is that these data are a very valuable part of the total amount of knowledge the developer used to create the product. In licensing and other commercial agreements, the licensees specifically pay for permission to use data that are regarded as confidential.

Expert scientific advice to regulatory authorities

Many regulatory systems use outside scientific experts on advisory committees to help assess a GM application. However, they only provide advice and do not make regulatory decisions themselves, as they are not accountable to the public. Those decisions should be the responsibility of the government regulatory authorities. These authorities should have a prime responsibility to protect public health and the environment. However, they should attempt to take into account all health and environmental issues, including the effect of refusing the permit. For instance, the decision by the Zambian authorities in 2002 and beyond to refuse food aid (see Trade issues, later in this chapter) may have contributed to problems of starvation.

Post-approval activities

A good regulatory system should continue to ensure human and environmental safety after the product enters the market. There should be post-approval monitoring to limit any adverse environmental or health effects that might arise. The system should also have authority to conduct inspections, sample food products, limit environmental problems should they arise and take legal actions against violators of permit conditions.

If all these regulatory processes are in place, a country should be able to use agricultural biotechnology with confidence. As commented by Per Pinstrup-Anderson, former Director General of the International Food Policy Research Institute (IFPRI), condemning agricultural biotechnology for its potential risks without considering the alternative risks of prolonging the human misery caused by hunger, malnutrition, and child death is as unwise and unethical as blindly pursuing this technology without taking into account the necessary biosafety regulations (Pinstrup-Andersen 1999).

Regulating transgenic crops sensibly

The above heading, from Bradford et al. (2005), reflects the argument that the costs of meeting regulatory requirements and market restrictions, although initially required, are now substantial impediments to the commercialisation of transgenic crops. Experience gained from long-accepted plant breeding methods, from two decades of research on and commercialisation of GM crops, and increasing knowledge of plant genomes indicate that if a specific trait encoded by a gene is safe, the genetic modification process itself presents little identified potential for lack of safety. They propose that as in conventional breeding, regulations should concentrate on phenotypic rather than genetic characteristics.

The costs and uncertainties that result from the rapidly proliferating national and international regulations are impeding development of new crops and new traits. The cost of meeting these requirements for major crops has been estimated at US$20 million to US$30 million per product (McElroy 2003). This limits commercialisation to a few multinational corporations and to traits in crops that have a large economic payback. Different genes and traits should be stratified into risk classes that would be subject to more appropriate regulatory requirements. For instance Bradford et al. (2005) suggest that the transgenic process should be deregulated and that regulatory classes be defined in proportion to their potential risks:

Deregulate the transgenic process

The phenotypes of GM plants and their safety should be the focus of regulations. The method used to produce them should, therefore, not be of concern. The product not the process used to produce it should be evaluated. After all, similar traits such as mutation-derived herbicide resistance developed through conventional breeding are not subjected to regulatory hurdles. In Canada for instance, a herbicide-tolerant crop is reviewed in the same way, regardless of whether it is obtained through conventional breeding or GM technology.

Create regulatory classes in proportion to potential risk

Regulations should treat classes of GM organisms differently based on the identified true risk associated with the traits. Reduced regulatory protocols should be associated with low-risk GM organisms where the new trait is equivalent to that developed by conventional breeding. Plant-derived pharmaceuticals (see Chapter 9) that have very low environmental toxicity, or those that are grown in non-food crops, are candidates for moderate risk regulations. Continued monitoring may be appropriate for plants with novel pest or herbicide-tolerance traits. Careful regulation of high-risk plants would be appropriate in cases where the transgene products have a documented possibility of causing significant harm to humans or the environment. Plants with the ability to accumulate heavy metals or other environmental toxins (see Chapter 9) could fall into this category.

Trade issues

In May 2003 the USA and its partners, Canada and Argentina, filed a suit with the WTO against the EU for allowing the de facto moratorium on GM crops to operate since 1998 (WTO 2006) The case numbers are DS291, DS292 and DS293, because the three countries filed separate submissions. Trade between the USA and Europe is considerable. In 2002 the USA exported agricultural products valued at US$6.1 billion, mainly grains and their by-products to the EU. The latter exported US$7.9 billion to the USA in mainly wine and alcoholic beverages (James 2003). Given that Bt and conventional maize are not routinely segregated in the USA, exports of maize to the EU have declined sharply, leading to an estimated loss of US$300 million per year. US maize exports to Europe accounted for 4% of its total agricultural exports in 1998. By 2002 they represented less than 0.1%.

This European opposition to GM crops comes despite declarations from the EU Commission following rigorous scientific assessment that GM food is as safe as conventional foods (James 2003). This has led many observers to conclude that the cause of the mistrust in GM products is political rather than scientific. Thus Austria, Luxembourg, Italy and Greece have banned GM crops. However, Spain, Germany, France and Portugal have commercialised small areas of Bt maize, with Spain growing it at 10% of its total crop. Of the new Eastern European accession countries Romania and Bulgaria have commercialised GM crops, while Hungary, Poland, Croatia, the Czech Republic and the Ukraine have conducted field trials.

In July 2003 the EU approved new and stricter regulations to label and trace GM crops that could lift the moratorium. However, the USA and its partners in the WTO suit have concluded that the proposed regulations are not workable, are not based on scientific assessment, and therefore violate WTO regulations. The EU's rationale for requiring stricter regulation is that labelling and traceability are essential to restore consumer confidence in regulatory oversight for food safety in Europe, required for lifting the moratorium. One of the US partnership's complaints about the regulations is that the consumers in Europe will perceive the labelling as a negative warning, which will lead to further losses of markets.

Another concern of many proponents of the value of GM crops for developing countries lies in the knock-on effect Europe's attitude could have on their use of GM crops. In particular, countries in Africa might opt to forego the significant advantages that GM crops offer because of the potential loss of export markets in Europe. For instance, as mentioned above, Zambia has refused maize, even milled maize, from the World Food Programme (WFP) since 2002 because it contains GM maize. This is despite widespread famine in that country that led to the looting of maize storage facilities by starving Zambians. Another example is Namibia where cattle farmers are afraid to feed their animals GM maize imported from South Africa in case they lose their lucrative European markets. However, the feed given to cattle would not be reflected in any potential labelling of the meat. The difference between the regulatory requirements of the EU and the buying strategies of EU distributors should be borne in mind. Some supermarket chains require their suppliers to provide them with totally non-GM products. This was the issue for the case of Namibia, not EU regulations.

A slightly different aspect on trade issues is the smuggling of GM soybean seeds from Argentina to Brazil (Bonalume 1999). Farmers in Brazil, where until recently GM crops could not be planted, realised that their counterparts in Argentina were reaping much improved harvests as a result of growing herbicide-tolerant GM soybeans. To solve the disparity, farmers smuggled seed across the border. However, the government has decided that farmers in Brazil may grow herbicide-tolerant soybeans legally. In addition Brazilians are also likely to plant Bt maize and Bt cotton on a commercial scale.

Another trade issue that was raised some years ago was whether countries could export beef that had been fed Bt maize to the EU. This was particularly of concern to cattle farmers in Namibia whose main exporting partner is the EU. However, the EU regulations EC 1829/2003 and EC 1830/2003 exclude meat products from animals fed GM feed or treated with GM medicines.

Intellectual property issues

Intellectual property (IP) can prove to be a major stumbling block in the development of GM crops, especially in developing countries. Ingo Potrykus, one of the developers of 'golden rice' was initially horrified to discover that this technology was subject to some 70 patents belonging to 32 different owners (Potrykus 2001) (see Chapter 9). However, he later concluded that if the companies who filed those patents had not done so, they would never have had the financial security to undertake their research and development. Thus, his work 'piggy backed' on the work protected by all those patents and enabled him to develop his product. In the event only six licences were required and licence fees were waived. This suggests that requests for waivers of licence fees to allow the use of patented technologies for the development of crops suitable for subsistence farmers may be possible.

In support of this possibility an organisation called the African Agricultural Technology Foundation (AATF) has been established (www.aatf-africa.org). Its aim is to facilitate the transfer of IP related to agricultural biotechnology, including GM crops, from multinational corporations to African seed companies and resource-poor farmers. In other words this is to allow Africans the freedom to operate using someone else's innovation. It is funded by, among others, the Rockefeller Foundation, the UK Department of International Development (DFID) and the US Agency for International Development (USAID). Its first breakthrough was a licensing agreement signed with Monsanto giving access to the *cry1Ab* gene for the development of insect-resistant cowpea in 48 countries in sub-Saharan Africa. The AATF can license it in Africa for use in cowpea and products made from the technology can be exported outside Africa for feed and food uses, but not for propagation. This protects African farmers and seed companies as only they will have access to the Bt cowpea. The AATF will own all improvements arising from their use of the technology and they agree to license such improvements to Monsanto on a non-exclusive, fully paid basis.

Summary

Different countries have different regulatory systems in place. The USA is the most experienced and has, to some extent, streamlined the application processes. The EU maintains a precautionary approach and tends to say 'no' or 'let's wait', rather than 'yes'. Most developing countries are in the process of establishing regulatory systems. The dispute between the USA, Canada and Argentina with the WTO against the EU has been settled. Patenting of processes or products may be a stumbling block in the development of GM crops, especially in developing countries but systems are being put in place to alleviate this. Time will tell how effective they will be.

References

Bonalume NR (1999) Smugglers aim to circumvent GM court ban in Brazil. *Nature* **402**, 344–345.

Bradford KJ, Van Deynze A, Gutterson N, Parrott W and Strauss SH (2005) Regulating transgenic crops sensibly, lessons from plant breeding, biotechnology and genomics. *Nature Biotechnology* **23**, 439–444.

Catchpole GS, Beckmann M, Enot DP, Mondhe M, Zywicki B, Taylor J, Hardy N, Smith A, King RD, Kell DB, Fiehn O and Draper J (2005) Hierarchical metabolomics demonstrates substantial compositional similarity between genetically modified and conventional potato crops. *Proceedings of the National Academy of Science of the USA* **102**, 14458–14462.

Chassy BM (2002) Food safety evaluation of crops produced through biotechnology. *Journal of the American College of Nutrition* **21**, 166S–173S.

Halford NG (2003) *Genetically Modified Crops.* pp. 61–64. (Imperial College Press: London.)

Jaffe G (2004) Regulating transgenic crops, a comparative analysis of different regulatory processes. *Transgenic Research* **13**, 5–19.

James C (2003) Global review of commercialized transgenic crops; 2002. ISAAA Brief no. 29. (International Service for the Acquisition of Agri-biotech Applications: Ithaca, NY.)

Lehesranta SJ, Davies HV, Shepherd LVT, Nunan N, McNicol JW, Auriola S, Koistinen KM, Suomalainen S, Kokko HI and Kärenlampi SO (2005) Comparison of tuber proteomes of potato varieties, landraces and genetically modified lines. *Plant Physiology* **138**, 1690–1699.

McElroy D (2003) Sustaining biotechnology through lean times. *Nature Biotechnology* **21**, 996–1002.

Pinstrup-Andersen P (1999) Biotech and the poor. *The Washington Post* 27 October 1999; A31.

Potrykus I (2001) Golden rice and beyond. *Plant Physiology* **125**, 1157–1161.

WTO (2006) World Trade Organization, Dispute settlement. http://www.wto.org/english/tratop_e/dispu_e/dispu_subjects_index_e.htm#gmo. [Verified 5 May 2006].

9
Future watch

Insect-resistant crops

In 1996 Bt crops were first commercialised. When they were introduced there was widespread concern that they would inevitably lead to insects developing resistance to the Bt toxin. However, the length of time that typically passes in the field before resistance to most conventional pesticides is found has already passed (Bates et al. 2005). Does this mean that the insect resistance management strategies put in place, such as the planting of refugia (see Chapter 2), have been effective? Or were the fears of resistance development unwarranted in the first place? It is too early to answer these questions but the occurrence of insect resistance to Bt toxins in glasshouse and laboratory experiments indicates that resistance in the field is a question of 'when' not 'if'.

A potentially effective method for delaying insect resistance is to deploy different Bt genes coding for different toxins on a rotation basis. When insects develop resistance to one toxin, plants producing the different toxin would be grown. Another method is to introduce both Bt genes into the same plant (i.e. stack the genes). If both toxins are expressed at high levels the frequency of insects with resistance to two toxins will be much rarer than resistance to one. In addition, crops that are affected by multiple pests may enjoy better protection by the deployment of two different toxins. Pests that are less susceptible to one toxin may be more susceptible to the second (as discussed in Chapter 2). Scientists tested the idea on broccoli plants under laboratory conditions where insects are able to develop resistance. They found that dual toxins significantly slowed down the emergence of insect resistance (Zhao et al. 2003). Such 'double-stacked' cotton plants have been introduced into Australia with considerable success (D. Tribe, pers. comm. 1995).

Another approach to preventing the development of insect resistance is to use genes that code for toxins different from Bt. The same bacterium, *Bacillus thuringiensis*, produces other insecticidal toxins that have different structures and functions. They also show insecticidal activity against a wide range of pests. Transgenic cotton producing one such toxin is being evaluated and commercial release is expected in 2008/09 in the USA (Bates et al. 2005).

Another approach has been to fuse the Bt toxin with the non-toxic region of the toxic ricin protein found in castor beans (see Chapter 2). This region binds to carbohydrate residues in the lining of insect guts and enhances the binding of the Bt toxin. This increases the effectiveness of the Bt toxin and also decreases the likelihood of insects evolving resistance as mutations would have to occur in the genes coding for both binding receptors simultaneously. Results have shown that transgenic rice and maize plants expressing this fusion protein are significantly more resistant to a range of insect pests than those expressing the Bt toxin alone. In addition, the fusion protein conferred resistance to a broader spectrum of insect pests (Mehlo et al. 2005).

Insect-resistant GM crops should not be seen as a 'magic bullet' for crop protection, however. Many traditional aspects of integrated pest management, such as biological control, crop rotation and the use of 'trap' crops to attract insects away from the main crop should not be ignored. For instance the International Centre for Insect Physiology and Ecology (ICIPE) in Nairobi has developed a 'push-pull' strategy based on the planting of companion crops to maize. Napier grass (*Pennisetum purpureum*) is a grass that attracts and traps stemborers (Khan and Mengech 2005). The grass is planted as a border around the maize fields where invading adult moths are attracted to chemicals emitted by the grasses. Napier grass can defend itself against the onslaught by secreting a sticky substance that traps the pest. The 'push' in the scheme is provided by a legume, *Desmodium* spp., which is planted between the rows of maize. The plants emit chemicals that repel the borers and drive them away from the maize crop.

Drugs in crops

Another type of GM crop that could be a life saver in Africa is plants engineered to produce vaccines. The WHO estimates that nearly three million people die annually of vaccine-preventable diseases. A large proportion of these people live in Africa. The diseases include measles, hepatitis, dengue fever, rotavirus-caused diarrhoea and poliomyelitis. The main reason for non-vaccination is the cost of vaccines. If these are not included in a state- or non-government organisation-sponsored vaccination program, the vaccines are simply too expensive for people to afford. There are other 'orphan' diseases, such as bilharzia, which affect people only in the tropics. The market for big pharmaceutical companies is too unprofitable to produce the relevant vaccines. Making vaccines in plants might be a way out of this blind alley.

Much has been written about the production of 'edible vaccines' (e.g. Fischer et al. 1999; Giddings et al. 2000; Thanavala et al. 2005; Santi et al. 2006). The initial idea was that pharmaceuticals, including vaccines, could be produced in a crop such as bananas. The naive idea was that people could be dosed by eating a transgenic banana. However, this is unlikely to work as it would be almost impossible to achieve consistently uniform doses. Different plants are likely to produce different amounts of the pharmaceutical and it is highly unlikely that any medical control council would sanction this approach. Later the idea arose that the food could be processed and sold as dried bananas or banana milk shake with the level of the pharmaceutical clearly indicted on the packaging. However, such is the controversy over this technology that one of the originators of the idea, Prof. Charles Arntzen, rues the day he ever coined the term 'edible vaccine'. Instead he uses the term 'plant-based vaccines'.

The question then is: which is the best host plant for producing pharmaceuticals? If, as some scientists advocate, that plant is maize or some other edible crop, pollen spread will have to be prevented. This could be done via one of the methods discussed earlier (see Chapter 6, Seed sterility), or by containment in a glasshouse. However, the costs involved will probably negate the advantage of cheaper production. A better solution might be to choose a non-food crop such as tobacco. This is an extremely hardy plant with very large leaves favouring high levels of production. It also has the advantage of converting a plant that can be made into a health-damaging product, into a health-provider. The protein to be used as a vaccine would be extracted from the plant and purified before being processed into a form suitable for oral application, such as a lozenge. It has been found that enough anthrax vaccine to inoculate the entire population of the United States could be grown on only one acre of tobacco when expressed in the chloroplast (Koya et al. 2005).

Why take this route instead of the 'traditional' vaccine production method of animal tissue culture or GM bacteria or yeast? The first reason is cost of production. Both these methods require expensive culture equipment. It is far cheaper to grow tobacco under controlled, biosafety compliant conditions. Second is safety. Pharmaceuticals produced in animal tissue culture can possibly become infected with animal viruses which, like HIV and Marburg/Ebola virus, can cross the animal–human barrier, unlike viruses such as Tobacco mosaic virus, a common tobacco pathogen, that is harmless in humans. Most plant pathogens are harmless to humans. Obvious exceptions are some of the fungi that infect plants such as maize and produce mycotoxins as discussed in Chapter 2. In the case of GM bacteria, many pharmaceutical proteins require chemical modifications after they are produced. These modifications can be done in plants and animals, but not in bacteria. Finally it is the cost of purification of the product and its administration. If a vaccine is to be administered intravenously it needs to be extremely pure and very stable. The major costs of a non-oral vaccine are therefore purification, keeping it cold and administering it by means of an injection. One of the reasons that smallpox was eradicated is that it occurred before the rise in awareness, as a result of the HIV/AIDS pandemic, of the transmission of viruses carried in body fluids. Sharing sterilized needles was not considered to be hazardous. This is obviously out of the question today, even after needle sterilization. The potential of this approach is shown by the success reported by Webster et al. (2005) in boosting a measles vaccine with an oral, plant-produced measles vaccine.

Consider therefore how plant-based vaccines would be purified and administered. When I was a girl we experienced a polio epidemic. We were lined up at school and a trained nurse gave each of us a cube of sugar into which a live, attenuated polio virus had been incorporated. By swallowing this we effectively became vaccinated as immunity was activated in the upper gastrointestinal tract, the route of polio transmission. Oral vaccines of this type can be used for viruses or bacteria that infect via the **mucous membranes** (i.e. via the upper respiratory or gastrointestinal tract – the mouth, pharynx, oesophagus and anus – as well as the penis or the vagina) (Streatfield 2006). So we are talking about HIV, severe acute respiratory syndrome (SARS) and the many third-world diseases that are transmitted via the **faecal–oral route** such as diarrhoea, dysentery, cholera and typhoid. As the vaccine is administered through the mouth and not intravenously, no needles are required. And finally, if it is formulated into a pill, the requirement for cold storage would be minimised.

A comparison of some of the features of pharmaceutical production in plants, bacteria and animal cell culture are shown in Table 9.1.

For developing countries the advantages are enormous. The plants could be grown locally in contained glasshouses and simple factories built nearby for the purification and formulation of the vaccines. These could then be supplied to local clinics for distribution under controlled conditions. Another production facility is being tested in the US state of Indiana – a former limestone mine. The scientists involved claim that they are achieving significant crop yields in a completely controlled environment (checkbiotech 2005).

Table 9.1 Comparison of some of the features of pharmaceutical production in plants, bacteria and animal cell culture[A]

	Transgenic plants	Bacteria	Animal cell culture
Risk	Unknown	Yes	Yes
Production costs	Low	Medium	High
Time effort	High	Low	High
Scale-up costs	Low	High	High
Product yields	High	Medium	Medium-high
Gene size	Not limited	Limited	Limited
Storage	Cheap/room temperature	Cheap/−20°C	Expensive/N_2
Distribution	Easy	Feasible	Difficult

[A] Adapted from Fischer et al. (1999).

What then is the down side to this approach? The major stumbling block is cross-pollination. It is not desirable that the gene encoding the pharmaceutical protein be transferred to its conventional crop counterpart, such as tobacco. Tobacco is not a food crop so the problem is not as great as if the gene might be transferred to a crop such as maize. The best way to prevent pollen transfer is to use the so-called 'terminator technology', or male sterility (see Chapter 6). This prevents the plant from producing viable pollen. In addition scientists have found that crosses between Nicotiana commercial tobacco and wild species produce completely sterile hybrid plants that look different from conventional tobacco (Zaitlin et al. 2004). However, even in the absence of male sterility, the tobacco producing the biopharmaceutical would have to be grown in biosafety compliant glasshouses, preferably far from any commercially grown tobacco.

How far have plant-based vaccines come? In a recent article (Thanavala et al. 2005) the authors report that potatoes genetically modified to carry a vaccine against hepatitis B, a virus that can cause liver cancer, liver failure and death, have been successfully tested on humans. More than 60% of the volunteers tested showed signs of increased immunity roughly equivalent to a standard vaccination after eating a single four-ounce portion (113 g) of raw transgenic potatoes. Charles Arntzen, the 'father' of plant-based vaccines, said that this could help combat a disease that infects 115 million people worldwide and kills more than one million per year (Kane 1995). There are already two conventional vaccines for hepatitis B, but high costs and distribution problems keep them from reaching more than half the people in under-developed countries.

Charles Arntzen's own team have reported that a vaccine against plague, caused by the bacterium *Yersinia pestis*, has been produced in tobacco. When the purified vaccine was administered to guinea pigs, the animals were protected against the plague. Plague is one of the oldest identifiable diseases and historically caused one of the largest numbers of deaths to humans. Commercially available vaccines are not very effective and have a range of side effects. This plant-based approach may usher in a cheaper and more effective preventative measure against this scourge (Santi et al. 2006).

Plants that clean up the environment

Lead is one of the most important heavy metal contaminants in industri-alised countries It has been called the number one environmental threat to the health of children in the United States, as stated by a US Environmental Protection Agency fact sheet (US EPA 2006). Cadmium contamination is also widespread, particularly in soils containing waste materials from zinc mines and in soils fertilised with cadmium-rich phosphate fertilisers. There is thus an urgent need to remove lead and cadmium from the environment. And that is the precise aim of a group of researchers in Korea and Europe (Song et al. 2003).

This process is called **phytoremediation**. Initial results, using genes from yeast, are very promising. For on-site application of transgenic plants the genes will have to be introduced into highly productive plants. The authors are developing poplar trees for this purpose.

A variant of phytoremediation is **phytovolatilisation**, whereby the contaminant is transformed and released into the atmosphere. This has been achieved for mercury but the performance of transgenic plants is not yet sufficient for commercialisation (Stewart 2004). Each case of phytovolatilisation would have to be checked that the contaminant cannot be converted in the atmosphere into a water-soluble compound that could be rained back onto the ground.

Another emerging use of GM crops is as **phytosensors**. These plants can be used to detect impurities in the environment, including landmines containing trinitrotoluene (TNT). Research into this has shown promising results in which plants, detecting TNT, are able to fluoresce and be detected from a distance using laser-based instrumentation. There is even progress in introducing genes into these plants that enable them to degrade TNT (Stewart 2004).

Transgenic plants resistant to other pests

Corn rootworm, caused by a group of insect species, is a particular problem in the USA causing losses estimated at more than US$1 billion per year (Mitchell et al. 2003). Larvae have caused damage generally only in fields where maize is grown consecutively. Thus, crop rotation has provided effective control. Recently, however, two variants of corn rootworm have emerged that make such rotations ineffective. Soybeans, often used in rotation, are now affected by a variant of corn rootworm that can attack both crops. In 2002, a telephone survey of 600 randomly selected farmers was conducted. Among the respondents, 92% treated their continuous maize and 62% treated their first-year maize with insecticide. Of these, 92% agreed that GM technology would be safer for humans than using soil insecticides and 82% agreed it would be safer for the environment. Scientists are working towards a corn rootworm resistant GM maize (Mitchell et al. 2003).

Genetically modified crops and hunger relief in developing countries

All the members of the United Nations (UN) have pledged to meet eight important development goals by the year 2015. The first is to halve the proportion of people who suffer from hunger and whose income is less than US$1 per day. Because more than 70% of the extreme poor who suffer from malnutrition live in rural areas, the effort to enhance agricultural productivity will be a key factor in achieving this goal and is listed as a key goal by the UN Hunger Task Force (Sanchez and Swaminathan 2005).

Some people will argue that the world produces enough food to feed its entire population, the only problem being redistribution. The flaw in this argument is that this solution has been around for decades and yet people in the developing world still go to bed hungry or die of starvation. In addition, most cattle and poultry are fed maize or soybean. These livestock require three to six times the amount of maize and soybean that could adequately feed the poor people of the world. Therefore, one possible conclusion is that, in order to feed these people, all maize or soybean should be used to alleviate food shortages. In the extreme, the production of meat, dairy products, eggs and poultry should be abandoned. This is clearly an impossible solution.

As for redistribution, political conflicts within and between countries present considerable hindrances. Transportation infrastructure is weak and costs of distribution are high. In addition, it is not always possible to take into account cultural preferences for certain types of food. For instance Kenyans prefer maize while Ugandans prefer bananas and plantains.

Some farmers in developing countries practise organic farming by default. This is not as the result of any philosophical choice but because they are unable to afford artificial fertilisers, insecticides and pesticides. Some people in developed countries view this situation with approval, believing that this is a 'natural' and desirable form of agriculture. They might be are unaware of the intensive inputs that are required by organic farmers in developed countries for continuous soil enrichment. Organic farming in developing countries can mean low input, and unsustainable farming practices. In this case, most crop yields are too low to provide leftover material to replenish the land, and the poor quality manure from their livestock is mostly burnt as fuel and not used as fertiliser, leading to soil degradation (Nuffield Council on Biotechnology 2004).

A substantial part of people's livelihood in developing countries depends on agriculture. Some estimates suggest as much as 70%. GM crops could provide these people with a substantial benefit. It is, therefore, unacceptable to reject the possible contributions that GM crops can make to reducing world hunger, malnutrition, unemployment and poverty.

It is sometimes argued that the introduction of GM crops into developing countries will adversely affect the informal seed sector whereby farmers keep or exchange harvested grain as seed for the next season. However, all currently available GM crops are hybrids and all farmers growing hybrid crops, whether they are GM or conventionally bred, have to buy hybrid seed every season in order to maintain the performance traits of those varieties. Nothing prevents farmers from retaining and re-sowing their own non-hybrid seed varieties or landraces if they wish to do so. But if new GM or hybrid seeds are preferred by farmers, because of their increased yields or decreased costs of chemicals, they will buy them. Even small-scale farmers are aware that open-pollinated crops, such as varieties of maize, give lower yields than hybrids. Thus, many farmers in developing countries have been buying hybrid seed from local or multinational companies for years (De Vries and Toenniessen 2001).

Crop yields in many African countries are much lower than those in the developed world. For instance, the average yield of a commercial maize farmer in the USA is 10 or more tonnes/ha, that of a commercial farmer in South Africa is about 4.5 tonnes/ha, and that of a small-scale farmer in sub-Saharan Africa is 0.045 tonnes/ha. There are various reasons for this. Soil acidity is one of them, with many African soils having a pH of 3.5 to 4.5. Not only does this require the addition of lime to neutralise the soil, it also results in the solubilisation of aluminium and manganese, which leads to toxicity. In addition, critical minerals such as molybdenum precipitate under acid conditions and are, therefore, not available to crops. Another problem in Africa is that the levels of phosphate in the soil are often low and sources of added phosphate are limited and expensive. Fertilisers are expensive. For example, the price of urea, a commonly used fertiliser, was US$400 per tonne in western Kenya, US$770 in Malawi and only US$90 in Europe in 2002 (Sanchez 2002).

In addition to these environmental problems, farming in developing countries is affected by pests, diseases and unfavourable climatic conditions such as drought. Although GM crops cannot address all these problems, protection against just a few could increase crop yields considerably.

Commercial farmers in developing countries are also enjoying the benefits of GM crops. In South Africa the national farmer of the year in 2005 said that GM maize is a boon to farmers. In 2005 he planted 300 hectares of insect-resistant Bt maize with excellent results. During the 2005/06 season he planned to increase his Bt maize hectarage.

Similarly in China, small-scale farmers planting Bt cotton reduced their use of pesticides by 60% to 80% between 2001 and 2002. The financial savings to 3.5 million farmers, who managed between 0.5 and 2 hectares, were considerable (Pray et al. 2002; Table 9.2).

Table 9.2 Average costs and returns in US$ per hectare for farmers surveyed in China, 2001

Cost	Bt	Non-Bt
Ouput revenue	1277	1154
Seed	78	18
Pesticides	78	186
Synthetic fertiliser	162	211
Organic fertiliser	44	53
Other costs	82	65
Labour	557	846
Total costs	**1000**	**1379**
Net revenue	277	−225

From Pray et al. (2002).

Similar improvements, including yield increases, have been achieved by small-scale farmers in South Africa. Farmers are members of associations that meet to provide support and discuss mutual concerns (Thirtle et al. 2003). In the 1999/2000 season, 12% of 1376 cotton farmers, who mostly managed farms of an average size of 1.7 hectares, planted Bt cotton. This rose to 60% the following season and is now more than 90%. Due to increased yields and reduced costs for pesticides, farmers were able to increase their gross margin significantly, despite nearly twice the cost of Bt cotton seeds compared with conventional seeds. The use of Bt cotton is reported to have also led to savings of about 1500 litres of water per farm (Nuffield Council on Biotechnology 2004).

One of the major achievements of the Green Revolution in the 1960s and 1970s was the development of semi-dwarf rice varieties. Shorter plants can make more nutrients available for grain production. Scientists have isolated a gene from a common weed, *Arabidopsis thaliana*, which codes for the same type of dwarfism found in the semi-dwarf wheat varieties used in the Green Revolution. When the gene was introduced into rice, dwarf plants were obtained (Peng et al. 1999). Researchers have introduced the gene into basmati rice which is usually tall, with weak stems susceptibile to wind and rain damage. Previous attempts to reduce the height of basmati rice using conventional means have been unsuccessful. Field trials with the GM varieties are underway. An important feature of this advance is in the conservation of biodiversity. The single gene can be inserted with minimal disturbance to the genotypes of locally well-adapted varieties.

Nutritionally enhanced transgenic crops

Researchers from Europe and the USA have developed a GM tomato with heightened levels of the nutrients lycopene and ß-carotene, the precursor to vitamin A, plus several other healthful compounds. Some of the nutrients have been increased by as much as 10-fold (Paine et al. 2005).

Vitamin A deficiency (VAD) is a common phenomenon in developing countries. In 1995, clinical VAD affected about 14 million children under five, of whom about 3 million suffered from xerophthalmia, the primary cause of childhood blindness (Nuffield Council on Biotechnology 2004). About 250 million children had subclinical deficiencies, increasing their risk of contracting ordinary infectious diseases. In many developing countries such diseases contribute significantly to high mortality rates, claimed by some as responsible for 3000 deaths per day and 500 000 cases of infant blindness per year (Anonymous 2005). At least one-third of the people who suffer from VAD deficiencies are found among the poor people of Asia who rely on rice as their staple crop and for whom alternative sources of vitamin A, such as green leafy vegetables, are not affordable.

Professor Ingo Potrykus and Dr Peter Beyer of the Swiss Federal Institute of Technology have developed a GM rice variety which produces enhanced levels of ß-carotene, the precursor of vitamin A or pro-vitamin A. The genes are derived from the daffodil (*Narcissus pseudonarcissus*) and a bacterium (*Erwinia uredovora*). The GM rice producing pro-vitamin A was crossed with another line containing high levels of available iron. Rice normally contains a molecule called phytate that ties up 95% of the iron, preventing its absorption in the gut. The GM rice contains a gene encoding an enzyme called phytase that breaks down phytate. As the resultant provitamin A/high available-iron hybrid rice is a delicate shade of yellow, the product has been called Golden Rice.

Golden Rice burst onto the international scene heralded by some as the biotechnology solution to VAD. However, even the best lines of the original Golden Rice accumulated ß-carotene to levels that supply only 15% to 20% of the recommended dietary allowance (RDA) of vitamin A. Although most people are only partly deficient in vitamin A and require only a small supplement to their daily diet, this was not acceptable. As a result the biotechnology company Syngenta went on to replace one of the daffodil's genes with a maize gene that resulted in 23-fold higher accumulation of ß-carotene (Paine et al. 2005).

Does that mean the company aims to monopolise its valuable product through exhorbitant licenses and sales of Golden Rice? Not at all. Syngenta is a member of the Humanitarian Golden Rice Network. The company will work with breeders in the public rice breeding institutions in Asia and South Africa to make locally adapted varieties of this Golden Rice 2 available to small-scale farmers with incomes less than

US$10 000. Once approved for release, varieties bred from Syngenta's rice will become the farmer's property, which can be sown year after year without payment (Anonymous 2005).

Unfortunately, the introduction of Golden Rice into target countries has been seriously delayed by the lengthy processes necessary to obtain permits to deploy seed for field testing. It will be interesting to see what the future holds for this extremely promising nutritionally-enhanced crop.

Golden sorghum and cassava are being developed using similar technologies (as discussed in the following text). Although Golden maize would be a nutritional boon for many poor people, especially in Africa, its public acceptance is debateable. To many Africans yellow maize is fed to animals and chickens, while only white maize is eaten by people. Whether it will be possible to introduce the genes encoding pro-vitamin A into maize without turning the kernels yellow will be the challenge for this crop.

New varieties of the so-called 'orphan' crops are also in the pipeline. These are crops that are of local importance but which are of no interest to major seed companies because of their small markets. They include cassava, chickpeas, pigeon peas, groundnuts, pearl millet and sorghum. In India, scientists at the International Crops Research Institute for the Semi-Arid Tropics (ICRISAT) are developing insect-resistant Bt chickpeas and pigeon peas and virus-resistant groundnuts (Randerson 2005).

GM crops are often thought of as products of multinational corporations, but in the developing world public research is spearheading the development of important new crop traits. Joel Cohen published an analysis of this work in 2005 (Cohen 2005). He highlighted the following:

In Africa, four countries (Egypt, Kenya, South Africa and Zimbabwe) are developing GM apples, cassava (see below), cotton, cowpea (see below), cucumber, grapes, lupin, maize, melons, pearl millet, potatoes, sorghum, soybeans, squash, strawberries, sugar cane, sweet potatoes, tomatoes, watermelons and wheat. The traits included agronomic properties, bacterial resistance, fungal resistance, herbicide tolerance, insect resistance, product quality and virus resistance.

Similarly in Asia, seven countries (China, India, Indonesia, Malaysia, Pakistan, Philippines and Thailand) are developing GM crops including bananas and plantains, cabbage, cacao, cassava, cauliflower, chickpeas, chilli, citrus, coffee, cotton, eggplant, groundnuts, maize, mangoes, melons, mung beans, mustard/rapeseed, palms, papayas, potatoes, rice, shallots, soybeans, sugar cane, sweet potatoes and tomatoes.

In Latin America, four countries (Argentina, Brazil, Costa Rica and Mexico) are developing GM lucerne, bananas and plantains, beans, citrus, maize, papayas, potatoes, rice, soybeans, strawberries, sunflowers and wheat.

There are several additional initiatives underway to improve local African crops. For instance, bananas and plantains are important food sources for a number of African countries. There are two major types of varieties used as staple foods: the east African highland bananas, which produce mainly cooking (matoke) and 'beer' bananas, and the plantains, mainly found in the lowlands of west and central Africa. The productivity of bananas and plantains in sub-Saharan Africa is severely constrained by pests and diseases including nematodes, banana bacterial wilt, banana weevils, the fungus *Fusarium* spp. and the bacterium known as black Sigatoka which causes leaf spot (AATF 2005). Laboratories within and outside Africa are working on solving some of these problems.

Florence Wambugu has spearheaded the use of tissue culture of improving banana production in Kenya (Wambugu 2001). The traditional way of propagating bananas is to uproot a young sucker from around the base of a mature plant. Although this is very cheap and easy, it has the disadvantage of transporting pests and diseases. Tissue culture breaks this cycle of infestation as it involves the production of fresh material under sterile conditions in a laboratory. The use of juvenile tissues and hormones in the culture media are further sources of plant vigour. In addition, when improved varieties, derived by conventional breeding, are propagated, the process of tissue culture leads to plants that mature earlier and yield more than conventionally propagated ones. The use of these plants can transform a smallholder's banana orchard from one that barely meets subsistence needs to a farm that provide produce to the markets of Kenya and beyond.

I visited some banana farmers near Nairobi in June 2004 and was extremely impressed by their yields of healthy fruit. I was told that the increased income generated by the new varieties had changed the status of the crop from being a 'woman's crop' to becoming a 'man's crop'. The difference is that 'women's crops' feed their families but 'men's crops' make money.

Cassava is a very hardy root crop that serves as a major subsistence staple in sub-Saharan Africa. It is also an important cash crop for many smallholder farmers in the region, and is key during famine. Unlike other major food crops, cassava is tolerant to poor soils and dry weather. The carbohydrate yield from cassava per unit of land is higher than for other major staples and it thrives across a wide range of ecological zones. It is normally available all year round, thus contributing to household food security. However, its roots, which represent its most economically valuable part, have poor keeping qualities and must be processed within three days (AATF 2005).

The Bill and Melinda Gates Foundation is funding the development of nutritionally-enhanced GM cassava under what is being called the 'BioCassava Plus' project. The cassava will have higher vitamin A and E levels. This is very important as cassava feeds 40% of Africa's population.

Deborah Delmer (2005) describes a meeting that brought together bench and field scientists at which breeders told molecular biologists that cassava flowers poorly and that two varieties needed for a cross often did not flower at the same time in the same breeding station. A project was started to create cassava that can be induced to flower at the appropriate time. This is important because cassava is very poor at flowering and two varieties one wants to cross often do not flower at the same time in the same breeding station.

Cowpea is the most important food grain legume in the dry savannas of tropical Africa, where it covers more than 12.5 million hectares. It is rich in high quality protein and contains almost as much energy by weight as cereal grains. Cowpea is consumed by nearly 200 million African. It provides cash income to smallholder farmers, serves as nutritional fodder for livestock, and provides an ideal way to complement protein-deficient diets. Unfortunately, cowpea productivity in traditional African farming systems is greatly reduced by biotic and abiotic stresses (AATF 2005).

Research is underway at various African and other research institutions under the umbrella of the Network for the Genetic Improvement of Cowpea in Africa (NGICA) to develop improved varieties of cowpea. The aims are to develop varieties that perform better in the face of these stresses, have higher yield potential, and have even greater nutritional value. One of the major insect pests is the *Maruca* pod borer. The AATF is negotiating with Monsanto for the use of the *cry1Ab*, whose product confers resistance to *Maruca*.

A very important initiative was recently announced – the Africa Biofortified Sorghum Project (ABS Consortium 2005), also dubbed 'The SuperSorghum Project'. It brings together nine institutions, seven in Africa and two in the USA, and is a truly public–private partnership aimed at improving the nutritional value of sorghum. It strengthens the recent report by the International Food Policy Research Institute (IFPRI 2005) that public research institutions in Africa are conducting 'groundbreaking' research on GM crops.

When considering the impacts that GM crops can have on developing countries it is useful to consider the following. There is a vast difference between farming practices on the fields of a farmer growing just one of two different crops on 500 hectares in the USA to another growing a range of crops on less than one hectare in Africa. The former will use hybrids developed from highly inbred lines and adapted to relevant climatic conditions. The latter, often a woman, will grow many different crops that minimise her risk if failure. For example she might plant some maize and beans in case rainfall is plentiful, and perhaps some sorghum, cassava and cowpea in case of drought. Cost considerations will prevent her from using even marginally acceptable levels of fertiliser or pesticides. These differences almost guarantee that any crop bred in the 'North' will not be adapted to her growing conditions (Delmer 2005).

In the same paper Delmer (2005) offers one solution to the improvement of agriculture in developing countries. She proposes that the model for public-private partnerships could be based on the development of

beneficial traits such as disease and pest resistance, or drought tolerance, in major crops perhaps being best addressed by the private sector,
whereas the public sector holds a wide range of locally adapted
germplasm relevant to poor farmers. In this way the public sector would
support efforts to transfer valuable private-sector traits/genes into
locally adapted varieties suitable for low-input agriculture. However, for
crops such as bananas, cassava, cowpea, sorghum and sweet potatoes,
mentioned above, the burden of crop improvement will fall to the public
sector, although the private sector should be strongly encouraged to find
ways to share relevant technologies and provide crucial advice.

As stressed throughout this book, GM crops are not the ultimate
solutions to improve agriculture worldwide. They are part of the solution and form an extremely important part of the solution, especially for
the poor in developing countries. However, if this technology is to
deliver the benefits outlined in this book, it will need broad public
acceptance and support. Everyone involved in its development will have
to become a more effective communicator as it is only through knowledge and understanding that the public will gain confidence in the use
of GM crops. This book is one scientist's effort to facilitate the understanding of the effects of GM crops on the environment.

Summary

New GM crops are in the pipeline. They include different insect-resistant plants, the production of pharmaceuticals, including vaccines in
plants, plants that clean up the environment, plants resistant to pests
other than insects, and with improved nutritional qualities, especially
vitamin A precursors. Improved crops of particular interest to developing countries are also in the pipeline. These include bananas, cassava,
cowpea, sorghum and sweet potatoes. All with an interest in sustainable
agriculture and food security for the poor should have their eyes open
for these improvements.

References

AATF (2005) A new bridge to sustainable agricultural development in Africa. Inaugural Report May 2002 – December 2004. (African Agricultural Technology Foundation: Nairobi, Kenya.)

ABS Consortium (2005) Africa Biofortified Sorghum (ABS) Project Consortium Fact Sheet http://www.supersorghum.org/fact_abs.htm [Verified 8 May 2006].

Anonymous (2005) Editorial. Reburnishing golden rice. *Nature Biotechnology* **23**, 395.

Bates SL, Zhao J-Z, Roush RT and Shelton AM (2005) Insect resistance management in GM crops, past, present and future. *Nature Biotechnology* **23**, 57–62.

checkbiotech (2005) Underground crops could be the future of pharming. http://www.checkbiotech.org/root/index.cfm?fuseaction=newsanddoc_id=10155andstart=1andcont [Verified 7 May 2006.]

Cohen, JI (2005) Poorer nations turn to publicly developed GM crops. *Nature Biotechnology* **23**, 27–33.

Delmer DP (2005) Agriculture in the developing world, connecting innovations in plant research to downstream applications. *Proceedings of the National Academies of Sciences of the USA* **102**, 15739–15746.

De Vries J and Toennissen G (2001) *Securing the Harvest, Biotechnology, Breeding and Seed Systems for African Crops.* (CABI Publishing: New York.)

Fischer R, Drossard J, Commandeur U, Schillberg S and Emans N (1999) Towards molecular farming in the future, moving from diagnostic protein and antibody production in microbes to plants. *Biotechnology and Applied Biochemistry* **30**, 101–108.

Giddings G, Allison G, Brooks D and Carter A (2000) Transgenic plants as factories for biopharmaceuticals. *Nature Biotechnology* **18**, 1151–1155.

IFPRI 2005 Public Biotech Crop Research in Africa. http://www.ifpri. org/media/20050707afbiotech.asp [Verified 7 May 2006].

James C (2005) Global status of commercialized Biotech/GM crops, 2005 ISAAA Brief 32. (ISAAA: Ithaca, New York.)

Kane M (1995) Global programme for control of hepatitis B infection. *Vaccine* **13**, S47–S49.

Khan ZR and Mengech AN (2005) Management matters in the war against stemborers. International Centre of Insect Physiology and Ecology. http://www.icipe.org [Verified 5 June 2006].

Kipruto arap Kirwa (2005) Kirwa backs adoption of technology to produce genetically modified crops. *The Nation*, 8 April 2005.

Koya V, Moayeri M, Leppla SH and Daniell H (2005) Plant-based vaccine: mice immunised with chloroplast-derived anthrax protective antigen survive anthrax lethal toxic challenge. *Infection and Immunity* **73**, 8266–8274.

Mehlo, L, Gahakwa D, Nghia PT, Loc NT, Capell T, Gatehouse JA, Gatehouse AMR and Christou P (2005) An alternative strategy for sustainable pest resistance in genetically enhanced crops. *Proceedings of the National Academies of Sciences of the USA* **102**, 7812–7816.

Mitchell P, Alston J, Hyde J and Marra M (2003) Benefits from transgenic maize resistant to corn rootworm. pp. 8–8. *ISB News Report* October 2003.

Nuffield Council on Biotechnology (2004) *The Use of Genetically Modified Crops in Developing Countries*. (Nuffield Council on Biotechnology: London.)

Paine JA, Shipton CA, Chaggar S, Howells RM, Kennedy MJ, Vernon G, Wright SY, Hinchliffe E, Adams, JL, Silverstone AL et al. (2005) Improving the nutritional value of Golden Rice through increased pro-vitamin A content. *Nature Biotechnology* **23**, 482–487.

Peng J, Richards DE, Hartley NM, Murphy GP, Devos KM, Flintham JE, Beales J, Fish LJ, Worland AJ, Pelica F et al. (1999) 'Green revolution' genes encode mutant gibberellin response modulators. *Nature* **400**, 256–261.

Pray CE, Huang J, Hu R and Rozelle S (2002) Five years of Bt cotton in China – the benefits continue. *Plant Journal* **31**, 423–430.

Randerson J (2005) By the people, for the people. Genetically modified crops don't just come from big corporations. *NewScientist*, **2487**, 36–37.

Sanchez PA (2002) Soil fertility and hunger in Africa. *Science* **295**, 2019–2020.

Sanchez PA and Swaminathan MS (2005) Cutting world hunger in half. *Science* **307**, 357–359.

Santi L, Giritch A, Roy CJ, Marillonnet S, Klimyuk V, Gleba Y, Webb R, Arntzen CJ and Mason HS (2006) Protection conferred by recombinant Yersinia pestis angigens produced by a rapid and highly scalable plant expression system. *Proceedings of the National Academies of Sciences of the USA* **103**, 861–866.

Song W-Y, Sohn EJ, Martinoia E, Lee YJ, Yang Y-Y, Jasinki M, Forestier C, Hwang I and Lee Y (2003) Engineering tolerance and accumulation of lead and cadmium in transgenic plants. *Nature Biotechnology* **21**, 914–919.

Stewart CN (2004) *Genetically Modified Planet*. (Oxford University Press: UK.)

Streatfield SJ (2006) Mucosal immunization – using recombinant plant-based oral vaccines. *Methods* **38**, 150–157.

Thanavala Y, Mahoney M, Pal S, Scott A, Richter L, Natarajan N, Goodwin P, Arntzen CJ and Mason HS 2005 Immunogenicity in humans of an edible vaccine for hepatitis B. *Proceedings of the National Academy of Sciences of the USA* **102**, 3378–3382.

Thirtle C, Piesse J and Jenkins L (2003) Can GM-technologies help the poor? The impact of Bt cotton in Makhathini Flats, KwaZulu-Natal. *World Development* **31**, 717–732.

US EPA (2006) Indoor air quality. http://www.epa.gov/iaq/lead.html [Verified 5 June 2006].

Wambugu FM (2001) *Modifying Africa: How Biotechnology can Benefit the Poor and Hungry, a Case Study in Kenya.* (Self-published by FM Wambugu: Nairobi, Kenya.)

Webster DE, Thomas MC, Huang Z and Wesselingh SL (2005) The development of a plant-based vaccine for measles. *Vaccine* **23**, 1859–1865.

Zaitlin D, Chambers OD, Li B, Mundell RE and Davies HM (2004) Drugs in crops. *Nature Biotechnology* **22**, 507.

Zhao J-H, Cao J, Li Y, Collins HL, Roush RT, Earle ED and Shelton AM (2003) Transgenic plants expressing two *Bacillus thuringiensis* toxins delay insect resistance evolution. *Nature Biotechnology* **21**, 1493–1497.

Glossary

Abiotic stress: stress conferred on a plant by non-biological phenomena such as drought, salinity, heat and cold.

Adventitious seed: the accidental presence of seed from another variety within a cultivar.

Aflotoxin: toxin produced by *Aspergillus* fungi that can cause oesophageal cancer and liver diseases in humans.

Amino acid: the building blocks of proteins.

Atrazine: a herbicide belonging to the triazine group that inhibits photosynthesis.

Back-cross: crossing hybrid offspring with the parent plant that is being improved.

Biodiversity: the variability among living organisms from all sources including, terrestrial, marine and other aquatic ecosystems and the ecological complexes of which they are part.

Biotic stress: stress conferred on a plant by biological organisms such as viruses, bacteria, fungi and insects.

Bt (Bt protein, Bt toxin): a protein toxin produced by the soil bacterium *Bacillus thuringiensis* which kills insect larvae. Bt toxins are not toxic to other animals or humans.

Chloroplast: plant organelle containing the green pigment, chlorophyll.

Cross-pollination: the transfer of pollen from the flower of one plant to the flower of another related plant. Successful pollination results in fertilisation and seed production.

DNA: deoxyribonucleic acid, the complex molecule that makes up the genetic material of an organism. It governs cell metabolism and is responsible for the transmission of genetic information from one generation to the next.

Ecosystem: a system made up of a community of animals, plants and bacteria and the physical and chemical environment with which it is interrelated.

Eutrophication: enrichment of nutrients in waterways, supporting increased growth of plants, algae and other waterborne biota. The decomposition of these organisms deprives the water of oxygen, thereby killing animal life.

Events: the insertion of a transgene in the genome of a particular plant cell that subsequently leads to a GM plant line with the transgene in a specific genetic location,

Faecal–oral route: infection that originates in faecal contamination of food.

Fumonisin: toxin produced by Fusarium fungi that can cause oesophageal cancer and liver diseases in humans.

Gene: the biological unit of inheritance that transmits hereditary information and controls the appearance of a physical, behavioural or biochemical trait. It is composed of DNA (deoxyribonucleic acid) and consists of a series of nucleotides (or bases) comprising its genetic code.

Genome: the total genetic composition of an organism.

Glufosinate: the active ingredient in the herbicide Liberty$^{®}$.

Glyphosate: the active ingredient in the herbicide Roundup.

Horizontal gene transfer: transfer of genes from one organism to another, often unrelated, species.

Hybrid: the offspring of two plants of different species or varieties.

Hybrid varieties: Crop varieties where specific male and female parents are crossed to produce a particular hybrid, with specific traits such as high yields and pest resistance. Hybrid seeds are bought by farmers every season.

Hybridisation: the production of offspring of two plants of different species or varieties.

Imazapyr: a herbicide used to coat the seeds of varieties of maize that have been bred to resist it. The herbicide kills *Striga* effectively.

Introgression: the insertion of the genes of one plant variety into the gene pool of another. This can occur when two varieties produce fertile hybrids. These can then be back-crossed with individuals of one of the parents.

Landrace: a variety of a crop that has been bred to perform well in a given area. It is selected under specific soil, climatic and pest conditions.

Mucous membranes: mucous-secreting tissue lining many body cavities.

Mutagen: an agent able to cause mutations.

Mutation: an inheritable change in DNA. The change is passed on from one generation to the next.

Mycotoxin: toxin produced by a fungus.

Open-pollinated varieties: crop varieties that have been bred so that farmers are able to plant their own seed produced by natural wind pollination for several seasons before having to buy seed. The yield is lower than that of hybrid varieties.

Organochlorine: a chlorine-containing insecticide of organic origin. Although very effective, they are non-selective (e.g. DDT) and most have been banned.

Organophosphate: a phosphate-containing insecticide of organic origin. Although some are still on the market, they are highly toxic to humans as they act as nerve inhibitors.

Phenotype: the visible characteristics of a plant.

Photosynthesis: process whereby plants use the energy of sunlight to synthesise carbohydrates from carbon dioxide and water.

Phytoremediation: use of plants to improve the environment.

Phytosensor: plants that can detect compounds, including impurities, in the environment.

Phytovolatilisation: the contaminant is transformed by the plant into a volatile compound that is released into the atmosphere.

Plant virus: an extremely tiny pathogen of plants that consists of a core of genes surrounded by a coat of protein.

Pyrethroid: a chemically synthesised insecticide mimicking pyrethrin which is obtained from the pyrethrum plant. They are relatively non-toxic to humans but insects rapidly develop resistance to them.

Refuge: Non-Bt varieties of a crop that allow susceptible insects to feed on them, survive, mate and compete with resistant insects that eat and emerge from Bt plants.

Rhizosphere: the air and soil directly adjacent to the roots of plants.

RNA: ribonucleic acid, as opposed to DNA (deoxyribonucleic acid), is the genetic material of some viruses.

Stolon: a creeping horizontal stem or runner that sprouts to produce new plants.

Striga: *S. hermonthica*, or witchweed, a parasitic weed that grows in the roots of plants such as maize, draining them of nutrients.

Transformation: the process whereby any gene can be introduced into a plant. Such a plant is then known as a transgenic plant.

Transgenic plant: a plant into which a gene from another species of plant, or another organism (e.g. a bacterium), has been transferred by genetic engineering.

Transposon: a mobile genetic element, consisting of a gene or series of genes, that can insert itself at random into an organism's genome.

Vertical gene transfer: transfer of genes from parent to offspring.

Volunteers: self-sown seeds derived from a previous crop.

Index